Studies
in the History of Mathematics and
Physical Sciences

3

Editors
M. J. Klein   G. J. Toomer

Irénée Jules Bienaymé
(Archives de l'Académie des Sciences de Paris)

C. C. Heyde   E. Seneta

I. J. Bienaymé
Statistical Theory Anticipated

Springer-Verlag
New York   Heidelberg   Berlin

C. C. Heyde

C.S.I.R.O. Division of Mathematics and Statistics, P.O. Box 1965, Canberra City, A.C.T. 2601, Australia

E. Seneta

Australian National University, Canberra, A.C.T. 2600, Australia

AMS Subject Classifications: 01A55, 01A70, 62-03

Library of Congress Cataloging in Publication Data

Heyde, C   C
    I. J. Bienaymé : statistical theory anticipated.

    (Studies in the history of mathematics and physical sciences ; 3)
    Bibliography: p.
    Includes indexes.
    1. Mathematical statistics—History. 2. Bienaymé, I. J., 1796–
1878. I. Seneta, Eugene, 1941–   joint author. II. Series.
QA276.15.H49   519.5′09   77-7367

All rights reserved

No part of this book may be translated or reproduced in any form without written permission from Springer-Verlag

© 1977 by Springer-Verlag New York Inc.

Softcover reprint of the hardcover 1st edition 1977

9 8 7 6 5 4 3 2 1

ISBN 978-1-4684-9471-6   ISBN 978-1-4684-9469-3 (eBook)
DOI 10.1007/978-1-4684-9469-3

. . . mais il faut parier. Cela n'est pas volontaire: vous êtes embarqué.

Blaise Pascal, *Pensées*

Preface

Our interest in I. J. Bienaymé was kindled by the discovery of his paper of 1845 on simple branching processes as a model for extinction of family names. In this work he announced the key criticality theorem 28 years before it was rediscovered in incomplete form by Galton and Watson (after whom the process was subsequently and erroneously named). Bienaymé was not an obscure figure in his time and he achieved a position of some eminence both as a civil servant and as an Academician. However, his name is no longer widely known. There has been some recognition of his work on least squares, and a gradually fading attribution in connection with the (Bienaymé–) Chebyshev inequality, but little more. In fact, he made substantial contributions to most of the significant problems of probability and statistics which were of contemporary interest, and interacted with the major figures of the period. We have, over a period of years, collected his traceable scientific work and many interesting features have come to light. The present monograph has resulted from an attempt to describe his work in its historical context. Earlier progress reports have appeared in Heyde and Seneta (1972, to be reprinted in *Studies in the History of Probability and Statistics,* Volume 2, Griffin, London; 1975; 1976).

It is our aim in this monograph to focus on Bienaymé's work in its context, both for its intrinsic interest and for the perspective it gives on developments in the 19th century. Indeed, the evolution of probability and statistics is fairly well documented up to the time of Laplace, and the developments of the twentieth century are widely appreciated. The intervening period of the last three quarters of the nineteenth century is the least well-understood period in the history of the subject. Reasons for this are not hard to find. The technical nature of the work of the period requires a substantial grounding in the subject on the part of its interpreters, as well as due caution in the imposition of modern interpretations. Furthermore, sources are frequently obscure enough to provide significant risks of the overlooking of important material and, at the same time, hindering early completion of most investigations. A work such as ours must inevitably be flawed by errors of omission and misinterpretation; of course we have tried to minimize them.

The timing of this book is opportune, since 1978 is the centenary of Bienaymé's death. In 1974 the centenary of the death of his close contemporary Quetelet was celebrated with considerable fanfare. A commemorative session was organized by the Royal Academy of Belgium in December 1974, and Belgium issued a commemorative portrait stamp at the same time. In addition, a group meeting at the 40th ISI Meeting in September 1975 was devoted to Quetelet. Centennial papers commemorating the contribution of Galton and Watson to branching processes (predated by that of Bienaymé) appeared in the journal *Advances in Applied Probability* in 1974.

There are various structural points concerning the monograph which should be remarked upon. First, quotations originally in French, German, and Russian are given in free English translation which seeks to capture the spirit rather than the letter of the original.

Next, there are several tables. Table 1 (in the front matter) lists the chronology of the most relevant scientists mentioned in the text; and Table 2 (at the back of the book) the references to Bienaymé extracted from the name indexes of *Comptes Rendus Hebd. des Séances de l'Académie des Sciences*. Bienaymé's publications are listed separately, rather than as part of the general bibliography.

In the case of a few individuals and books, we have not been able to positively ascertain full details on items such as initials, place of publication, or publisher. Also, a number of papers and books mentioned have been reprinted. In these cases we have sought to give at least one bibliographically complete reference, while mentioning others.

Finally, in cases where not only multiple, but confusing, versions of names exist, generally due to transliteration, we have used the following: Bortkiewicz, Chebyshev, Chuprov, Hanikov, Iastremsky, Liapounov, Sleshinsky, even though these authors may have themselves used different transliterations (which are indicated in the reference list).

The extant memorabilia of I. J. Bienaymé which we have been able to trace consist of the following:

(i) A dossier on I. J. Bienaymé in the Archive of the Académie des Sciences, Paris, which contains, *inter alia,* a photograph of I. J. B. (reproduced as our frontispiece), a note concerning the location of his tomb at the Montparnasse cemetery, a decree concerning his election to the Academy, and certain documents pertaining to his death and that of his wife (which are discussed in our Chapter 1).

(ii) Three letters from I. J. B. to Chebyshev, in the Archive of the Academy of Sciences of the U.S.S.R. These are reprinted in Russian translation in Chebyshev (1951) and are discussed in our Chapter 1.

(iii) A photograph of M. (= Monsieur) Bienaymé preserved in the *Département des Estampes de la Bibliothèque Nationale,* but without initials or date.

Finally, acknowledgments are due, first, to O. B. Sheynin for invaluable assistance with materials; and also to P. Aukland, C. Eisenhart, R. A. Horváth, P. Jagers, H. O. Lancaster, D. R. McNeil, H. Seal, H. Solomon, and S. Stigler. The assistance of an Australian Research Grants Committee Grant, the Australian National University Library, the French Embassy in Australia, the Bibliothèque Nationale, the Académie des Sciences de l'Institut de France, and the Statistics Department at the Virginia Polytechnic Institute and State University, is also gratefully acknowledged. In addition, our thanks are due to Ms. H. Patrikka for typing the manuscript; and last, but not least, to our wives and families for their forbearance.

C. C. Heyde
E. Seneta

Canberra, A.C.T. and
Blacksburg, VA
March, 1977

Table of contents

Table 1. Chronology of most relevant scientists mentioned in the text

(by year of birth and nationality)

French

A. Deparcieux	1703–1768
"Moheau"*	1733–1820
J. A. N. de Caritat de Condorcet	1743–1794
P. S. de Laplace	1749–1827
A. Legendre	1752–1833
E. E. Duvillard	1755–1832
J. Fourier	1768–1830
L. F. Benoiston de Châteauneuf	1776–1856
S. D. Poisson	1781–1840
D. F. Arago	1786–1853
A. L. Cauchy	1789–1857
G. Lamé	1795–1870
I. J. Bienaymé	1796–1878
A. A. Cournot	1801–1877
J. Liouville	1809–1882
C. Hermite	1822–1901
J. L. F. Bertrand	1842–1912

Other

R. Price	1723–1791
T. R. Malthus	1766–1834
T. Doubleday	1790–1870
F. Galton	1822–1911
H. W. Watson	1827–1903

*Pen name of J. B. A. R. Auget, Baron de Montyon.

C. F. Gauss	1777–1855
W. Lexis	1837–1914
L. A. J. Quetelet	1796–1874
M. V. Ostrogradsky	1801–1862
V. A. Buniakovsky	1804–1889
P. L. Chebyshev	1821–1894
A. A. Markov	1856–1922
L. J. Bortkiewicz	1868–1931
A. A. Chuprov	1874–1926
A. P. de Candolle	1806–1893

1. Historical background

There is properly no history; only biography.

R. W. Emerson

1.1. Introduction

Up to the 1820s and including the work of Laplace, developments in probability and statistics were largely inseparable. Scholars who worked in the area had broad scientific interests and official, as distinct from mathematical, statistics had remained in its comparative infancy. The history up to this time is fairly well understood, in large part through Isaac Todhunter's book (1865), which gives an encyclopedic treatment. See also Hacking (1975) and David (1962), which essentially do not go past Moivre; and Maistrov (1974). For the less mathematical aspects of statistics see, for example, Westergaard (1932) and Meitzen (1891).

From the early 1830s the subject of statistics, then principally understood as data collection, quickly rose to importance. At the forefront was the need for government official statistics for planning purposes. Official statistics began to diverge from mathematical statistics at the outset; and from roughly the beginning of the twentieth century, probability and mathematical statistics also began to develop along different pathways. The profound developments which took place in statistics were principally led by Pearson, and later Fisher, in the U.K.

Pearson, originally an applied mathematician and philosopher of science, was strongly influenced by the biologists Galton and Weldon and started to work in statistics from 1890 or 1891. He had little awareness of non-British antecedents of his work, antecedents which, indeed, were often in rather obscure places. Furthermore, the evolution of the subject over the first quarter of the twentieth century was so rapid as to lead to inevitable difficulty over attributions, ideas being ascribed to Pearson and Fisher, for example, which had earlier origins. Nevertheless, what has happened in statistics since the time of Pearson is fairly well understood.

The history of the intervening period of nearly 100 years from Laplace to Pearson is not well understood or well documented. With the exception of discussions on narrow specialized areas, and the possible exception of the

Soviet contributions (e.g., see Maistrov, 1974), there is principally a range
of rather restricted biographical material (e.g., *Dictionary of Scientific
Biography* and sources referenced therein) and little material on the
detailed evolution of the subject. I. J. Bienaymé [1796–1878] was an
important, but hitherto largely ignored, representative of this period. He
made substantial contributions to the significant problems which were of
contemporary interest and interacted with the major figures of the period. It
is the object of this monograph to focus on his work, both for its intrinsic
interest and for the light it throws on developments in the nineteenth
century. Bienaymé was not an obscure figure; he achieved a position of
some eminence both as a civil servant and as an Academician. The fact that
comparatively little of his work has been afforded due recognition is a
reflection on the past and current historical perspectives of his period.

1.2. A historical prelude

To establish the context for the later discussion we shall give a brief
historical perspective.

The effective birth of probability as a subject took place in hardly more
than a decade around 1660. There is comparatively little history to record
before this time, although the emergent ideas of probability were recognized earlier, for example, in the sixteenth century writings of Cardano.
The work of, and correspondence between, Pascal and Fermat was a vital
component in the birth of the subject. After having been introduced to this
work on a visit to France in 1655, Huygens published the first probability
text in 1657. At about this time, Pascal applied decision–theoretic reasoning to the problem of the existence of God and applied probabilistic
reasoning for the first time to contexts other than one of games of chance.[1]
Graunt used the London Bills of Mortality as a basis for the first extensive
set of inferences on death rates, which he published in 1662. Leibniz (1665)
established an important connection between evidence and probability and
the first serious attempt to put annuities on a sound actuarial basis was
published by de Witt in 1671.

The posthumous publication of J. Bernoulli's book *Ars Conjectandi*
(1713) was the next major milestone. This book provided a significant study
of combinatorial games of chance, subsuming the work of his predecessors.
It also contained a rigorous proof of the simplest form of the weak law of
large numbers, the first limit theorem of probability theory. This theorem
deals with a system in which repeated independent trials are possible and
there is a constant probability p of "success" at any trial. If S_n is the
number of "successes" obtained in n trials the result states that $Pr(|n^{-1}S_n - p| < \epsilon) \to 1$ as $n \to \infty$ for any $\epsilon > 0$. This opened up, for the first time, the

---

[1]See §5.8.

possibility of wide application of probability theory to statistics. However, Bernoulli found only crude limits on the probability in question and this issue, in particular, was taken up in the major work of Moivre (1718). Here the simplest ($p = \frac{1}{2}$) case of the Central Limit Theorem:

$$\lim_{n \to \infty} Pr(a < (np^{-1}(1-p)^{-1})^{1/2}(n^{-1}S_n - p) < b)$$
$$= (2\pi)^{-1/2} \int_a^b e^{-u^2/2} \, du, \quad -\infty < a < b < \infty,$$

was obtained. Laplace later extended this result to the case of general p, but even this, for the important case of unknown p, does not allow the computation of the practically important quantity $Pr(|n^{-1}S_n - p| < \epsilon)$. This issue was taken up in the posthumously published work of Bayes (1763), in which a careful solution was given to the problem of finding the posterior distribution of p given S_n when the prior distribution of p is uniform on the interval [0,1].

Up to this time, the calculus of probabilities had been largely restricted to games of chance and actuarial problems. Wider applications had been little more than touched upon. Now the subject matter broadened out and came solidly to bear upon problems of philosophy, and upon political and social science, notably in the study of demography and of the procedures of representative bodies and judicial panels. A significant role in stimulating this transformation was played by Condorcet, especially in his book (1785). However, it was Laplace who gave the transformation real mathematical substance and placed French probability and statistics firmly in the ascendancy.

In his classical treatise (1812, with new editions in 1814 and 1820) Laplace presented the basic results known in probability theory. He systematized and extended the previously known theory and laid the foundations for applications in many new areas. This was a culmination of his work and relatively little of the content or method of the treatise was new at the time of publication. Most of the material consisted of a generalized reworking of his papers dating from as early as 1771.

Laplace was an innovator as well as a synthesizer. He introduced the theory of generating functions and the basic ideas of the theory of characteristic functions as a tool to use in the context of the Central Limit Theorem. He developed what is now usually known as Bayes's Theorem to serve as a basis for statistical inference. Also, he made major contributions to the theory of errors and the Method of Least Squares. Here his framework was that of a linear regression in one explanatory variable with independent and identically distributed residuals possessing a symmetric density. He utilized asymptotic methods to justify the use of least squares, in contrast to Gauss (1809), who initially employed a fixed sample size and normally distributed residuals. Both used methods which today would be considered as decision–theoretic.[2]

---

[2]See §§4.1–4.2.

The field of demography was one with which Laplace was especially concerned. He discussed methods of constructing mortality tables and for making indirect population counts, his plan for the latter being used to estimate the population of France in 1802. He extended the earlier work of D. Bernoulli on the influence of disease on mortality rates and the problem of inoculation against infectious disease. He developed confidence intervals for the proportion of male births in a population.

Laplace improved much earlier work on classical problems, for example, that of partitioning the stakes in an interrupted series of games. He extended J. Bernoulli's Weak Law of Large Numbers and Moivre's Central Limit result rigorously to the case of discrete lattice random variables with finite support and less precisely to the general case of identically distributed random variables. Also, following Condorcet, he worked on the probabilistic analysis of jury decision making, work that later led to a change in the French jury system.

Laplace has sometimes been criticized for applying the theory of probability more widely than was claimed appropriate but he had a balanced attitude toward applications and warned (1774, p. 645):

> . . . that the science of chance must be employed cautiously and requires modification when passing from the mathematical to the physical case.

Apart from Laplace there were other highly significant figures on the French mathematical scene in the 1820s, in particular A. L. Cauchy [1789–1857], J. B. J. Fourier [1768–1830], and S. D. Poisson [1781–1840]. The primary interests of these three individuals were in fields other than probability and statistics but they each made major contributions to the area and they enter into the account given in this monograph.

The vast mathematical work of Cauchy, which appeared over a period of almost 50 years, had a profound influence ranging over much of pure and applied mathematics. In particular, his classic *Cours d' analyse* (1821) set new standards of rigor. Unfortunately, his character was flawed by bigotry and an extremely strong desire to display his intellectual superiority over others. He contributed to probability and statistics only through his work on the theory of errors, in which context he was involved in a major controversy with Bienaymé.[3]

Fourier's major areas of mathematical work were in heat diffusion, partial differential equations, and the theory of equations. His contributions to probability and statistics were principally in the area of errors of measurement from large numbers of observations. These were published in reports of the Bureau of Statistics of the Department of the Seine, of which he was Director, over the period 1821–1829. Some of this material was later to motivate a major piece of work of Bienaymé's.[4]

---

[3]Discussed in detail in §§4.4–4.7.
[4]See §§3.5 and 5.6.

Finally, Poisson's major mathematical interests were in mechanics, mathematical analysis, and physics. He was a vigorous rival of Cauchy and Fourier, particularly in the area of heat diffusion.[5] It was toward the end of his life, roughly from 1835 on, that Poisson's interests began to focus on probability and statistics. His major work in the area, essentially a treatise on probability limit theory with applications to juridicial statistics, appeared in 1837 (Poisson, 1837b). Bienaymé was later to dispute various results in this work, particularly a version of the Law of Large Numbers.[6]

It is against this background that the later work of the nineteenth century in probability and statistics, and in particular that of Bienaymé, must be considered. Much of Bienaymé's work was concerned with extending or defending Laplacian positions, for example, in connection with least squares, judicial statistics, and insurance. He also worked along established lines in the areas of demography and social statistics. He did not lack innovative ability, however, and produced in particular the so-called Bienaymé–Chebyshev Inequality 15 years before Chebyshev, a recognition of the concept now known as sufficiency 80 years before its formalization by Fisher and the Criticality Theorem for branching processes 28 years before its incomplete rediscovery by Galton and Watson.

1.3. Biography[7]

Irenée Jules Bienaymé was born in Paris on 28 August, 1796, and died there on 19 October, 1878 (many accounts give the date as 20 October), at age 82. His life therefore coincided with a stormy period of French history.

---

[5]An interesting commentary on the work of these men and their rivalry is given in Grattan-Guiness (1972). Chapter 22.

[6]See §§3.3 and 2.4.

[7]The biographic material of this section (§1.3) is taken from three principal sources written close to the death of Bienaymé. The first is the obituary note presented to the Academy of Science, Paris, by J. de la Gournerie (1878), who was closely academically associated with Bienaymé in both the Academy (a part of the *Institut de France*) and the *Société Mathématique de France*. The second is the biographical notice by A. Gatine appearing in Volume 3 of the three centenary volumes, published in 1894–1897, of the *École Polytechnique;* this adds only a little to the note of Gournerie. The third is the encyclopaedia entry by L. Sagnet (undated). It will be seen that Bienaymé spent many years of his life in the civil service: Sagnet is listed in the encyclopaedia as *"attaché au Ministère des travaux publics,"* and so is (possibly as a colleague) in a good position to give information on this period.

Other useful sources are Vapereau (1880) and Poggendorf (1898).

Two more recent sources utilized are the biographic notes of Franceschini (1954) and Dugué (1968); only brief information is contained in the *Grande Dictionnaire Universel du XIXe Siècle* of Larousse.

The dossier on Bienaymé held in the archives of the Academy of Sciences, Paris, has provided information on his death–interment documents and also a list of his surviving blood and law relations, with their positions and honors.

The French Revolution began in 1789, and Louis XVI and Marie Antoinette were guillotined in 1793; the *École Polytechnique* was founded in 1797. Napoleon seized power late in 1799 and was deposed in 1814, to be followed by the Bourbon Restoration. The revolution of 1830 resulted in the replacement of Charles X by Louis Phillipe as king, and so in the termination of the older branch of the Bourbons as rulers. The latter reigned until forced to abdicate in 1848. The revolution of that year resulted in the setting up of the Second Republic, which lasted till 1852, when Louis Napoleon (to become later Napoleon III) seized full power and initiated the Second Empire [1852–1870].

Bienaymé began his secondary education at the *lycée* in Bruges (Belgium was then part of the French Empire) and completed it at the *Lycée Louis-le-Grand,* Paris. In 1814 (at age 18), he took part in the defence of Paris in a company of volunteers, in the face of the advancing allies. He enrolled in the *École Polytechnique* in 1815; this institution was dissolved in 1816, since its students persisted in their loyalty to the Napoleonic regime.[8] Subsequently he worked as a translator for journals. In 1818 Bienaymé became lecturer in mathematics at the military academy at St. Cyr, leaving in 1820 to enter the Administration of Finances. He soon became Inspector and became Inspector General in 1834 (the revolution of 1830 did not interrupt his civil service career). In 1844 he became an officer of the Legion of Honor. On account of the revolution of 1848, he retired at this time (age approximately 52). Economic conditions had been bad and Public Works had been organized to relieve unemployment; but these only provided the focus for further agitation. Thereafter, Bienaymé devoted all his time to scientific work with the exception of a hitherto unknown and very rare publication (Bienaymé, 1851) on the alignment of houses, which takes the form of an extensive satirical dialog and displays a literary bent (Gournerie mentions membership of literary circles), as well as a known abhorrence (Sagnet) to the structure of the law.[9]

So far as can be determined, his first published memoir is *Mortalité des nouveaux-nés* (Bienaymé, 1829), which appears when he was aged approximately 33. He became active in the affairs of the *Société Philomatique* (later *Philomathique*) *de Paris,* an association for the advancement of science. His many contributions to its *Procès Verbaux* begin to appear in print in 1837 in the scientific newspaper–journal *L'Institut, Paris,* and are each reprinted, usually at the end of the year of their appearance, in the compendium *Procès-Verbaux de la Société Philomatique de Paris.*[10] One of these (Bienaymé, 1840d) contains a clear recognition of the statistical notion, later observed to be of major importance, of a sufficient statistic;[11]

---

[8]cf. Bradley (1976).
[9]See §6.2.
[10]See §1.6.
[11]See §5.6.

the last (Bienaymé, 1845) contains the Criticality Theorem for branching processes.[12] In the period prior to his retirement, he also presented in 1834 and 1835, respectively, two substantial memoirs (Bienaymé, 1838b, 1837a, respectively) whose publication was somewhat delayed. His own comments on the contents and dates of his publications until 1852 are given in Bienaymé (1852b).

Shortly after his retirement in 1848, he was temporarily appointed professor at the Sorbonne in the Calculus of Probabilities. The academician Lamé, who later became holder of the chair (and whose own published contributions to probability and statistics appear few), spoke of him in the following glowing terms on 26 April, 1851, at a resumption of the Course in the Calculus of Probabilities:

> It is my pleasure to count amongst my friends a savant (M. Bienaymé) who today, almost alone in France, represents the theory of probabilities, which he has cultivated with a kind of passion, and in which he has successively attacked and destroyed errors. It is to his counsels that I owe a proper understanding of the true scope of the science which I teach, and of the limits beyond which it cannot pass without losing its way.
> (Reported in *Nouvelles Annales de Mathématiques,* for June, 1851)

The year 1852 saw the publication of one of his most famous memoirs (1852a) in Liouville's journal, on the probabilities of errors in the Method of Least Squares. The memoir was also presented to the Academy, and we shall speak later[13] regarding a very favorable report by Lamé, Chasles, and Liouville. On account of this and other interesting memoirs, according to Gatine (1897), this same year saw Bienaymé's election to the Academy of Science, on 5 July, at age 55. These various steps may be traced from Table 2. It is perhaps not without interest that the *coup d'état* that inaugurated the Second Empire (of Napoleon III) was *de facto* effected in 1851, and we have noted strong evidence that Bienaymé was a Napoleonic sympathizer.

Indeed it is noted by Gournerie, and repeated by Gatine and Dugué, that despite his retirement Bienaymé had considerable influence as a statistical expert in the government of Napoleon III. He was attached for 2 years to the Ministry of Commerce. In a report to the Senate on 26 April, 1864, the Minister of Commerce, M. Dumas, spoke of help given during his ministry with the organization of the government retirement fund by:

> . . . an honorable member of the Academy of Science, M. Bienaymé, whose competence in these matters all Europe knows.

Gournerie goes on to say that the information given by Dumas showed that the tariffs calculated by Bienaymé had assured an almost perfect balance in

---

[12]See §5.9.
[13]See §4.3.

the operations of the fund. We may note that Bienaymé's early work concerns what we may now call social statistics and demography and displays an interest in the yet earlier work of Châteauneuf, Deparcieux, and Duvillard.

Bienaymé's entry to the Academy was soon followed by publication of his best known memoir (1853c), an impassioned defence of the Laplacian position in regard to linear least squares, which contains *interalia* both the Bienaymé Equality and the Bienaymé–Chebyshev Inequality.[14] This occurs in the context of a famous controversy with the great mathematician Cauchy[15] in this general subject area, which began almost immediately on Bienaymé's election. Shortly after, we find Bienaymé (1855) criticizing, invalidly, the Law of Large Numbers of Poisson,[16] on which subject Bienaymé had presented a memoir 13 years before, in 1842, to the *Société Philomatique* but which had not been published *"par hasard"* according to his own description (1852b). It is noteworthy that Poisson died in 1840. Dugué (1968) remarks:

> Scientific controversies had considerable appeal for Bienaymé.

We shall explore this further in the sequel and mention at present only that he was also involved, somewhat later, in a substantial controversy over actuarial matters.

During his time as academician, he acted for 23 years as a referee (numerous times with Gournerie) for the Prize for Statistics of the Montyon Foundation, the highest French award in this area. The collection of his reports on this and other matters published in the *Comptes Rendus* form a considerable body of work on their own (see Table 2). He was a founding member of the *Société Mathématique de France;* Council member [1872–1878], Vice President [1874], and President [1875]. He was made a life member in 1875. This period saw the publication in 1874 of his last writings in the society's *Bulletin,* on a remarkable limit theorem, again far ahead of its time, pertaining to the number of local maxima and minima in a series of random observations.[17]

At the time of his death he was, additionally, corresponding member of the St. Petersburg Science Academy and the Central Commission of Statistics of Belgium (his association with Chebyshev and Quetelet is discussed in the following section) and honorary member of the Association of Chemical Conferences of Naples. He had also been awarded the medal of S$^{te.}$ Hélène.[18]

---

[14]See §5.10.
[15]See §4.6.
[16]See §3.3.
[17]See §5.11.
[18]Instituted in 1857, the medal has on one side the effigy of Napoleon, and on the other, in French, "Campaigns of 1792 to 1815. To his comrades in glory, his last thought. St. Helena, 5 May, 1821."

Bienaymé had a considerable knowledge of European languages; according to Gatine he was an early student, among the French intelligentsia, of Russian and, indeed, translated an article of Chebyshev (1855) for Liouville's journal. He knew the classical languages well; in his 1851 satire, we find some (presumably original) verse in Latin. In 1870 he presented to the Academy an explanation of two passages of Stobée pertaining to the mathematical knowledge of the Pythagoreans, which had hitherto remained a mystery. His knowledge of classical Greek is further manifested in a large-scale project which his death interrupted: an annotated translation of Aristotle, of which several parts were completed in manuscript. According to Sagnet:

> He possessed a universal erudition, having studied all branches of human knowledge. . . .

It is evident from three surviving letters from Bienaymé to Chebyshev (Chebyshev, 1951), that in the last 5 years of his life he was in poor health and unable to correspond. He had difficulty in sleeping and often suffered from some form of trembling fits. During 1875, when he was president of the *Société Mathématique de France,* he was unable to attend any of the sessions. He was buried at Montparnasse cemetery on 22 October, 1878.

We have been able to obtain rather little information on Bienaymé's family and personal life. His wife was (née) Françoise Gabrielle Clémence Harmand who died on 8 October, 1876, aged 72 years and having received the sacraments of the church.[19] Their address at the time was *Rue de Fleurus* No. 1, Paris. Their son was Arthur François Alphonse Bienaymé, born in Paris on 13 January, 1834, elected correspondent for the section of geography and navigation of the Academy of Sciences on 29 January, 1900, who died at Toulon on 25 January, 1906. He left the *École Polytechnique* in 1853 and entered the marine engineers. Later he campaigned in China in 1860–1863 in charge of a gunboat. In 1871 he was made member of the *commission permanente d'examen des mèchaniciens de la flotte.* In the letter of 24 December, 1874, from Bienaymé to Chebyshev, he is mentioned as a captain of the engineering corps. He was appointed head of *l'école d'application du génie maritime* at Cherbourg in 1880 and then in Paris [1882] where he taught the course on steam engines. He was Director of Naval Construction at Toulon, became *Inspecteur Général du Génie Maritime,* and published a book in the area (A. F. A. Bienaymé, 1887), shortly before his election. He also had one son (at least), Colonel Bienaymé de la Motte. Another son of I. J. Bienaymé may have been L. Irenée Alexis Bienaymé, born 8 April, 1836, who left the *École Polytechnique* in 1857, and in 1878 was *"Chef du Bataillon du Génie";* and there is evidence of younger children.

---

[19]This information is derived from Academy records and it seems significant that no sacraments are mentioned in connection with I. J. Bienaymé.

There seems to have been a general family connection with the Ministry of Finances. A relation, Charles Phillipe-Aimé Bienaymé, who died on 27 June, 1870, at age 70, is described in the Academy records as *"ancien Chef de Bureau au Ministère des Finances."* On I. J. Bienaymé's wife's side there is M. Alphonse Harmand, *"Inspecteur général des Finances."* Furthermore, an examination of the general catalog of holdings of the *Bibliothèque Nationale* reveals a number of works written in about 1900 on demographic and social-statistical topics by one Gustave Bienaymé; a relation with this name was *"Chef de Bureau au Ministère des Finances"* in 1878. He is probably not a son of I. J. Bienaymé, but a more distant relative. Nepotism does seem a possibility.

1.4. Academic background and contemporaries[20]

We have noted that Bienaymé's first paper appears at a relatively late age, and apart from (the summaries of) his contributions to the proceedings of the *Société Philomatique* he wrote little. A generous assessment might put the total number of papers at about 22. His own list of publications (1852b), presumably connected with his election to the Academy that year, lists 14 items.

We should note at the outset, and shall return to the point repeatedly, that the treatise of Laplace (1812) appears to have been the guiding light of much of his work. Some of the best work, particularly on linear least squares, is concerned with elaborating, generalizing, or defending Laplacian positions; and the whole body of his work makes it evident that he is (albeit at a temporal distance) a true and loyal disciple of Laplace. Although it is likely that Bienaymé was essentially self-taught, it is conceivable that he actually came in contact with Laplace during the one year [1815–1816] at the *École Polytechnique*. Laplace was actually appointed first full professor of this institution on its foundation in 1797. The memoir of David (1965) suggests that there were more than a few negative aspects of Laplace's ethics; but then Bienaymé himself was, evidently, not totally in accord with the accepted behavior patterns of his times. For recent very positive assessments of Laplace's contributions in probability and mathematical statistics, see Stigler (1975a) and Sheynin (1974). A detailed commentary on Laplace's life and times is given by Crosland (1967).

It is quite possible that Bienaymé was also significantly influenced by Fourier, who certainly provided the motivation for *Probabilités* (Bienaymé, 1840d), one of his most important papers (see §5.6). Fourier was appointed as Director of the Bureau of Statistics of the Department of the Seine (which included Paris) in 1815 and held the post until his death in

---

[20]Consultation of the dates in Table 1 will facilitate the reading of this section.

1830. He would presumably have had dealings with the Ministry of Finance and possibly Bienaymé. Fourier certainly influenced Quetelet, about whom we shall shortly have more to say. Indeed, Quetelet, on a number of occasions, quoted Fourier's statement:

> Statistics will make progress only as it is retained in the hands of those versed in higher mathematics.

as a guiding principle [e.g., Hankins (1908), pp. 15–20, and sources cited therein]. For a recent biography of Fourier see Grattan-Guiness (1972).

Bienaymé's early papers (1829, 1835, 1837a, b, 1838a, 1839b), in demography and social statistics as already noted, display in the last three mentioned a strong probabilistic component. He was later to return to this area in a group of papers (1857, 1862a, b, c, 1865), all concerning insurance statistics, the theme of an earlier paper (1839b). It is known (Darmois, 1928), that he was a correspondent of Quetelet, the Belgian Astronomer-Royal and the leading exponent in his day of such analysis—see Quetelet (1835, 1844, 1846, 1848)—among other things (Hankins, 1908). Indeed, Bienaymé is mentioned in Quetelet (1848, p. 306) in regard to the Law of Large Numbers as a distinguished mathematician whose works have clarified several difficulties in probability theory. Bienaymé's membership in the Central Commission of Statistics of Belgium is complemented by Quetelet's membership of the *Société Philomatique*. A curious fact is that the two men were born the same year and died within a few years of each other. There has recently been a revival of interest in the work of Quetelet associated with the centenary of his death (e.g., Horváth, 1973, 1975; Godeaux, 1973; Lévy, 1975; Stigler, 1975b; and a *Mémorial Adolphe Quetelet,* Académie Royale, Bruxelles, 1975). According to Horváth, there are substantial reasons for calling him "the father of statistics." Certainly his stimulus played an important part in the formation of the London Statistical Society (in 1834), now the Royal Statistical Society; also, the initiative leading to the establishment of international statistical congresses, the first held in Brussels in 1853, came principally from Quetelet. Also, Horváth is of the opinion that his sociometric contributions need modern reappraisal and that his contribution to mathematical statistics/probability is underappreciated. Quetelet is the central figure in the period 1830–1849, marking the "era of enthusiasm" in statistics.

In the same area, it may be that, apart from his acquaintance with the work of Châteauneuf (1824), Bienaymé was familiar with the work of Candolle, whose later book (1873) is associated with Galton's and Watson's (Galton, 1873; Watson, 1873; Galton and Watson, 1874) investigation of the extinction of surnames, which had in fact been completely resolved much earlier by Bienaymé (1845). Bienaymé's solution may have been obtained under the inspiration of an almost simultaneous (but uncited) investigation by Châteauneuf (1845), and a slightly earlier one by Doubleday (1842), on noble families, preceding both Galton's and Candolle's.[21] It is curious that Bienaymé is not mentioned by Candolle, in spite of the book's title, although Galton is mentioned in the 1873 edition in connection with his *Hereditary Genius* (1869); and both Galton and Watson certainly are in the 1885 edition. Galton's own later book (1889) essentially reproduces Galton and Watson (1874) in an Appendix but has no mention of Bienaymé, although it is evident from the 1892 edition of *Hereditary Genius* that Galton was well aware of the work of both Cournot and Quetelet, close associates of Bienaymé.

Antoine Augustin Cournot [see the encyclopaedia entries of Feller (1961), Guitton (1968), and Nogar (1967); the article of René (1933); and the books of Harpe (1936), Mentré (1908, 1927), Milhaud (1927), and Segond (1911) listed] was a man of very great intellect, whose breadth of interests and, particularly, capacity for writing were enormous. In this breadth, and in other respects, his career and its historical role parallel those of Bienaymé. Thus, most of his life was spent in university administration (at one stage he was Inspector General of Studies); he was a linguist and translator of several mathematical books. In parallel to Laplace with Bienaymé, Poisson was strongly influential in Cournot's career; and Cournot, like Bienaymé, married and had a son. Still in close association with Bienaymé, his work was criticized after his death by Bertrand (1883) and he has disappeared almost entirely from the literature of probability. [Bertrand criticizes Bienaymé in a later book (1889).]

Cournot wrote ten books between 1838 and 1877: these have three major themes: algebra, infinitesimal calculus, and calculus of probabilities; the theory of wealth; and philosophy of science, philosophy of history, and general philosophy. Although he is sometimes mentioned as the founder of mathematical economics and retains a place in this context as well as in a general philosophical one, as with probability he has almost disappeared from the philosophy of science. He is said to have been modest and self-effacing, a devout Catholic; the following assessment by Guitton (1968) perhaps is partly true also of Bienaymé:

> Cournot was a pioneer. He did nothing to court his contemporaries, and they, in turn, not only failed to appreciate him, but ignored him. By a fitting reversal, his triumph came 80 years after his death. The most advanced of the econometric school recognize him as their ancestor.

Cournot's (1843) book on probability, largely philosophical and notable for an early "frequentist" view of the subject as later developed by von Mises—see the summary by O. Anderson (1954, pp. 131–132)—mentions[22]

---

[21]See §5.9.
[22]See §3.2.

two of Bienaymé's (necessarily early) papers in the text and carries the following conclusion to its Preface (p. VII):

> I shall not conclude this preface, which is perhaps too long, without expressing here my debt to my excellent friend, M. Bienaymé, inspector general of finances, whose work in statistics and probability is well known to mathematicians. Long occupied without each other's knowledge with the same objects of study, we have found ourselves united by singular conformity of ideas and tastes. He has kindly undertaken to advise me regarding the printing of my book, even so far as reviewing the proofs and reworking some of the numerical calculations. He has done this with an impartiality all the more remarkable since he had long ago arrived at a theory of posterior probabilities from premises quite different from those which have guided me, and this theory appears to me to revert in essence to that to be found in Chapter VIII of the present work. The reader will be able to judge the similarities and differences if he decides to publish his own researches, which, I hasten to stress, will still be entirely original, and in which his skill in the use of analysis will doubtless enable him to find many things which have escaped me.

Bienaymé's loyalty to Cournot, as well as the bitterness which apparently existed between the former (an agnostic?) and Cauchy (a bigoted Catholic), is manifested even well after Cauchy's death, in a communication to the Academy (Bienaymé, 1871). Cournot had published in the *Bulletin des Sciences Mathématiques, Physiques, Chimiques* (of Ferussac) a number of articles under the initials A. C., although his custom in general was to use his full surname; six of these were attributed to Cauchy (who always used his full surname) in the first volume of the *Catalogue of Scientific Papers* of the Royal Society of London, even though Cournot's full surname had appeared in the corresponding list of authors of the *Bulletin*. The same error occurs in the biohagiography of Cauchy by Valson (1868), as Bienaymé pointed out, and is perpetuated yet again in the February 1869 volume of the *Bolletino di Bibliografia e di Storia delle Scienze matematiche e fisiche*, Rome. In these two sources, the fault lies in a table of Cauchy's publications prepared by one Narducci. In this the articles are not even presented in chronological order (as in the English catalog) but are grouped under appropriate journal headings, with the result that under the *Bulletin* heading, Cauchy's and Cournot's articles are thoroughly confused.

Elements of the early acquaintance between Bienaymé and the great Russian mathematician Chebyshev, alluded to in the three surviving letters,[23] dates back at least to 1858, which saw the appearance of Bienaymé's translation, with a long prefatory footnote, of Chebyshev's (1855) paper, and certainly to 1867, which saw Chebyshev's article of that date juxta-

---

[23]See §1.3. These were written in 1874–1875; it may be relevant to note that the inaugural congress of the Universal Postal Union took place in 1874.

posed, in French translation, to a reprinting of Bienaymé's article (1853c) in Liouville's journal. This translation (as also of several other papers) was by one N. V. Hanikov, a Russian living in Paris, whom the second of the three letters (24 December, 1874) reveals to have been a close and valued friend and intermediary of both men.

The first of the letters (27 May, 1874) expresses Bienaymé's gratitude for Chebyshev's magnanimity in his complimentary references to the "method of Bienaymé" in Chebyshev (1874), read at a Congress at Lyons in August 1873 but which had just come to Bienaymé's attention. Another point mentioned in this letter is the French Academy's recent election (18 May, 1874) of Chebyshev as *associé étranger,* which, we may presume, came about largely through Bienaymé's efforts. At the time of the second letter (24 December, 1874), Bienaymé had just been unofficially informed by Chebyshev, through the agency of Hanikov, of election to the St. Petersburg Academy as corresponding member. In this reply he expresses his great joy and gratitude for the honor and his strong suspicion that, indeed, the election came about largely through the efforts of Chebyshev as referee. The third letter (14 August, 1875) refers to a recent trip to Paris by Chebyshev; it was at this time, during Bienaymé's presidency, that he addressed the *Société Mathématique de France.*

The general preamble to Chebyshev's correspondence in Chebyshev (1951), reveals a number of other facts about him which are relevant to this account, as also to the sequel. Indeed, Chebyshev was a notoriously bad correspondent; his closest French colleague was Ch. Hermite, who had acted as intermediary between Chebyshev and the ill Cauchy. Hermite had been elected corresponding member of the St. Petersburg Academy in 1858 and was still corresponding with Chebyshev in 1890. Their names are bracketed for the so-called Hermite–Chebyshev polynomials, though Hermite appears to have rediscovered them later.

Bienaymé's other, and main, area of contribution was in probability theory and mathematical statistics (Bienaymé, 1838b, 1839a, 1840a–1852, 1853c, 1874, 1875). Insofar as he is remembered at all today, it is in this area: for the inequality that bears his name and Chebyshev's, and for his work and contributions toward the understanding of linear least squares. We shall discuss in the sequel a number of contributions (some already mentioned) of equal importance. These were too far ahead of their time; correct, but generally without proof in published form, and couched in extensive verbiage because of Bienaymé's preoccupation with applications. They are now almost completely forgotten.

Doubtless historically detrimental to him also were the facts that he left no disciples, not being an academic; that he wrote no book—and it is largely on their books rather than their papers that the fame of earlier mathematicians seems to rest; and his controversy with Cauchy and his attack on Poisson. The blistering attack on his views on least squares in the influential book of Bertrand (1889, §226) would not have helped[24]; in this connection it is noted by Sleshinsky (1892) that Bertrand is also antipathetic to Chebyshev, whom he does not deign to mention.

Even in the inequality, which Bienaymé obtained by the simple proof still used today[25] in contrast to Chebyshev, Chebyshev's name takes precedence in general usage, for such reasons as:

> . . . it was first clearly stated and proved by Chebyshev . . .

according to the Russian historian of probability Maistrov (1967, 1974).

In addition to the two major areas, Bienaymé published, as befits his career, a few discussions on actuarial matters (see Table 2),[26] and apart from an article on the history of mathematics already mentioned (Bienaymé, 1870), another (Bienaymé, 1843) which casts new light on Pascal's contribution to the foundation of the probability calculus.[27]

1.5. Bienaymé in the literature

We have remarked already that reference to Bienaymé in the scientific literature is scant; he suffers considerably in comparison with his contemporaries mentioned in the preceding.

There appear to be only a few works devoted to the history of probability. The earliest, that of Gouraud (1848), antedates some of his contributions and, not altogether surprisingly, does not mention him. The best known, that of Todhunter (1865), essentially covers the history only to Laplace but mentions Bienaymé at two points (Sections 1041 and 1043),[28] insofar as his work relates to Laplace's *Théorie Analytique,* notably evidencing Bienaymé's devotion to Laplace's authority, in the face of criticism by Poisson. The other authoritative reference on the history of probability, the extensive and very good survey of Czuber (1899) lists four of Bienaymé's most important contributions, describing especially his role in the development of linear least-squares theory and his turning-point test and limit theorem, but none of those to the *Société Philomatique.* A sequel, a list by Wölffing (1899), refers to these last, at least for 1840, and mentions also Mansion's (1881) note on the turning-point test. The bibliographical

---

[24]See §4.3. Another such attack concerned judicial statistics, to which Bienaymé had also contributed: see §2.4.

[25]See §5.10.

[26]Discussed in §2.5.

[27]Discussed in §5.8.

[28]See §§2.4, 2.5, 3.2.

work of Merriman (1877) mentions, in all, 9 of Bienaymé's papers, to wit, those of 1838b*; 1839a, where the pagination is confused with 1839b; 1840b, with wrong pagination; 1840d; 1840e, with wrong pagination; 1852a*; 1853a*; 1853c*; and 1875*. The asterisks indicate the presence of a brief commentary; there is a general inability to penetrate to the essence of Bienaymé's work.

Of the less known works, the statistical (rather than probabilistic) discussions of John (1884)—which extends only to 1835—and Meitzen (1891) and the Hungarian probabilistic sources of Láng (1913) and Jordan (1972) omit all mention, although Bienaymé is mentioned once by Horváth (1967). There is only one mention in passing (and with no citation) in the history of Maistrov (1967), of which the English edition (1974) by S. Kotz is more accurate (see pp. 200–202).

Additionally, there exists a number of sources, which are actually treatises on probability but contain much historical material, that are often useful in historical investigation. The work of Laurent (1873), appearing at the end of Bienaymé's life, is disappointing in only listing nine of his papers (a listing reproduced more recently by Keynes (1921)). The book of Meyer (1874, 1879) contains an extensive account of aspects[29] of Bienaymé's linear least-squares theory, while the nonparametric test is mentioned again in Savage's (1962) bibliography. The treatise of Mises (1964) contains two citations. The more specialized book of Lancaster (1969) also contains several references in its early pages.

Recently a number of articles of historical nature have appeared which throw increasingly more light, at this late stage, on his activities. We mention Lancaster (1966), Seal (1967), Stigler (1974), and Harter (1974a), as well as our own articles, which together are intended to summarize the present book: Heyde and Seneta (1972, 1975, 1976). A surprisingly good, though old and obscure, article with an extensive discussion of some of Bienaymé's contributions is that of Sleshinsky (1892), which is mentioned a number of times in the present work.

Although in various encyclopedia entries we find the following nutshell descriptions of Bienaymé: *publiciste français, statisticien et administrateur francais, le savant algébriste*, and Gatine's biography appears in a section entitled *Carrières Financières*, there is an absence of more than fleeting references in French literature, both probabilistic and not. Although appearing at a favorable time, the work of Bachelier (1914) contains no mention, as is the case with his other books. Likewise, Pasquier (1926), while emphasizing Cournot, omits Bienaymé. In demography, a definitive work, such as that of Levasseur (1889), mentions him only in connection with the Law of Large Numbers, save for a reference to a Montyon Prize report which he delivered to the Academy of Sciences in 1872. Even in a

---

[29]See §4.3.

self-congratulatory piece on the *École Polytechnique*, such as Callot's (1958), he figures only in some tables.

Outside of France, in the social sciences, such works as the *Handwörterbuch der Sozialwissenschaften* accord generous coverage to Cournot and Quetelet but none to Bienaymé; and he is unfortunately omitted from his correct alphabetical place in the *Dictionary of Scientific Biography*.

More pleasing is the fact that, even though Bienaymé's contribution to combinatorics is small—it is contained in Bienaymé (1874, 1875)—he is nevertheless mentioned by Netto (1901).

1.6. The *Société Philomatique* and the journal *L'Institut*

The *Société Philomatique de Paris* (*Philomatique* became *Philomathique* officially *circa* 1851) was founded in 1788. A comprehensive historical statement concerning it and its publications is given by Berthelot (1888) on the occasion of its centenary. We give only a few details, mainly to cast light on Bienaymé's membership and his frequent contributions to its *Séances Verbaux*.

There were several such societies at this time, such as the *Société Phil. de Bordeaux,* founded 1808, and the *Société Phil. de Verdun (Meuse),* founded in 1822. Their aim appears to have been the advancement of science. The Paris society, among other publications, issued a *Bulletin,* 1807–1918, whose frequency of publication and title varied; e.g., 1825–1833, *Nouveau Bulletin des Sciences;* 1848–1861, *Extraits des Procès-Verbaux des Séances.* From 1836 to 1875 the proceedings of this society (among others) were printed in *L'Institut, Journal Universel des Sciences et des Sociétés Savantes en France et l'étranger,* and were reissued annually or quarterly under the title *Extraits des Procès-Verbaux des Séances* and *Bulletin* [1864]. We thus find that the articles of Bienaymé, active in the society between 1837 and 1845, are printed twice: first in *L'Institut* and then in the *Extraits* of the same (or next) year. It is relevant to note that *L'Institut* [1833–1865], beginning with 1836, was issued in two sections; Bienaymé's talks were reported in Section 1—*Sciences mathématiques, physiques et naturelles*—issued weekly (monthly July 1837– April 1838).

A decline in the frequent publication as an autonomous journal of the *Bulletin* in 1833 was due to the introduction in 1835 by Arago of the illustrious *Comptes Rendus Hebdomadaires des Séances de l'Académie.* The quick and easy publication of memoirs presented to the Academy rendered "useless the partial and abbreviated publication by the *Société Philomathique*" (Berthelot, 1888). However, when *L'Institut* became defunct, more frequent publication was to recommence. After several further name variations (e.g., *Bulletin de l'Association Français pour l'Avancement des Sciences,* at a conference of which association at La Rochelle Chebyshev played a prominent role in 1882), it became again, in 1953, the *Revue Générale des Sciences Pures et Appliquées,* under which name it continues to this day.

The *Société Philomat(h)ique* has included many famous French savants. Poisson and Fourier both had periods as editor of the *Bulletin,* Poisson serving for more than a decade before the time of Bienaymé (elected 17 January, 1838). We find, for example, Lamarck, Biot, Poisson (elected 1804), Gay-Lussac, Ampère, Arago, Dulong, Cauchy (elected 1814), Fresnel, Becquerel, Savart, and Coriolis, in chronological order. In the article of Berthelot (1888), where the composition of the society is given as of 26 October, 1878, Bienaymé is mentioned as the last deceased member, with date of death given as 19 October, 1878. At this time, Bertrand was also a member.

In conclusion to this section, it is relevant to note that because of the double publication of Bienaymé's articles in *L'Institut* (articles generally titled) and in the *Extraits* of the *Société Philomatique* (articles generally untitled), there has been confusion over not only the titling, but over the distinction between articles themselves. Thus in Kendall and Doig's (1968) bibliography, the foundation of our investigations, there are six articles listed as distinct, with distinct titles, which in actual fact amount to only three articles; this can be deduced from Bienaymé's own publication list (1852b) and, of course, from the articles themselves. We hope to have given a correct listing in the present work, although we have not been able to examine all the *L'Institut* versions published.

2. Demography and social statistics[1]

Nor is the Peoples Judgment always
true:
The Most may err as grosly as the Few.

J. Dryden

2.1. Introduction

Although Bienaymé is now chiefly noted for his contributions to probability and mathematical statistics, he entered the area via the subject of demographic statistics. This route of entry is primarily a consequence of some 28 years spent as a civil servant in the Administration of Finances, where his attention would presumably have been focused on day-to-day problems in the area of official statistics. Additionally, as noted in §1.3, he rather lacked formal mathematical training but might have been influenced by Fourier, who occupied the post of director of the Bureau of Statistics of the department of the Seine from 1815 until his death in 1830.

Developments in official statistics were considerable during the time period in question. By the 1830s the importance of detailed statistics for planning and organizational purposes was generally acknowledged. This transformation of attitude had taken place recently, although the setbacks had been considerable in the early 1800s. The *Bureau de la Statistique Générale* was established in France in 1800 but suppressed in 1812. The *Société de Statistique de France* was founded in 1803 and disappeared in 1806, while the statistical journal, *Annales de Statistique,* created under Government auspices and first published in 1802, lasted only 2 years. It was in the late 1820s that the perspective changed. Annual statistical publications by the Ministry of Justice began in 1827. The *Annales d'Hygiène publique* founded in 1829 published many statistical articles from its inception. The *Bureau de la Statistique Générale* was reestablished in 1833. The census improved significantly in reliability and information content, and while that of 1831 was not particularly satisfactory, the one of 1836 was decidedly better and designated the inhabitants by family and household for

---

[1]See also §6.3.

the first time. Official publications on economic statistics, on imports and exports, shipping and agriculture, began to appear in the late 1830s and early 1840s. A statistical bureau in the Ministry of Public Works was created in 1844. Bienaymé worked through this era of comparative statistical enthusiasm. We recall from §1.3 that he joined the Administration of Finances in 1820, becoming Inspector General in 1834, a post that he held until 1848.

In other European countries the statistical developments of the time were roughly parallel to those in France. In England, for example, a statistical department was added to the Board of Trade in 1833, and 1837 saw the establishment of civil registration of vital statistics. Statistical Societies were founded in Manchester in 1833 and London (later to become the Royal Statistical Society) in 1834. England and France had, in fact, both attempted a general census for the first time in 1801.

It is interesting to note early opposition to the collection and publication of official statistics, even in countries with a free constitution and relative stability throughout the period. In 1753 a Bill to provide for an annual census was introduced into Parliament in England, but it aroused a storm of protest and was rejected by the House of Lords. Possibly the controversy following on the publication of Thomas Malthus's *An Essay on the Principle of Population* (1798) was partly responsible for the passage of the Bill of 1801, leading to the first decennial census. Enumerations dating from some 50 years earlier were made in Scandinavian countries but were often not published, such information sometimes being regarded as a State secret or there being no official organization to adequately process the results. Nevertheless, the waning and later waxing of activity corresponded to that in England and France. Part of the difficulty had been through pressure from clergy, who had been charged with the collection of substantially increasing amounts of information and had reacted strongly against this.

Interesting historical accounts of the early development of statistics, particularly official statistics and demography but excluding mathematical statistics, are given by Westergaard (1932), by Meitzen (1891), and in *The History of Statistics,* collected and edited by J. Koren (1918), where various authors detail developments in their own country. Even more detailed accounts, with similar emphasis, are given by Levasseur (1889), Mols (1954), and Glass (1973). Unfortunately there are no such comprehensive treatments in the area of mathematical statistics, as distinct from probability theory.

In this chapter we trace Bienaymé's contributions in the area of demographic and social statistics. These are taken by subject in chronological order of first contribution and cover Infant Mortality and Birth Statistics, Life Tables, Legal Statistics, and Insurance and Retirement Funds. Bienaymé began to publish in 1829, but some of the contributions discussed in this chapter occur quite late in his career and indicate that, although he had gradually turned to the more mathematical areas of statistics, he had not lost interest in the applied areas whence his motivation must have originated.

2.2. Infant mortality and birth statistics

Bienaymé's publications begin in April 1829 with what is essentially a letter to the editor of the *Moniteur Universel*.[2] This concerns the mortality of newborn children, motivated by a report appearing approximately a week earlier in the newspaper *Globe* of a lecture on the same subject given to the Royal Academy of Sciences on 6 April. This lecture was given by a M. Fontenelle, who had quoted data collected in Castel-Franco in Italy by a Dr. Trévison. It was asserted on the basis of this data that of 100 children born in winter, only 19 would survive to age 1, while of 100 children born in spring, summer, and autumn, the numbers surviving to age 1 would be 48, 83, and 58, respectively. Bienaymé was struck by what he called this frightening infant mortality and its implications and proceeded to make an analysis, leading to the conclusion that the results must be incorrect.

Bienaymé took, from the book *Recherches sur la population des Pays-Bas* by M. Guételet, data of a Dr. Lastri on the number of births per season from the registers in Florence over a 323-year period, to compute a figure for the proportions of births per season in Italy. Then, subjecting these proportions to the mortality rates cited above, he observed that the mortality of infants up to age 1 would be very close to 50 in 100—that is, very near the mortality rates for foundling children. He quotes the extensive investigations of Châteauneuf (1824) and the typical European figure of 25 in 100 (the highest of Châteauneuf's figures being 45 in 100 for Vienna). He then comments on the implications of the quoted data with regard to birth dates of the mature age population and, in the absence of appropriate Italian data, observes the distribution of birth dates per season in French mortality data of 1818 and 1819, which shows little seasonal influence. While observing the data to be highly questionable, his conclusions were mild. His principal concern was to:

. . . question the effect of the winter . . .

and he was happy to conclude that:

Undoubtedly Providence, in a mother's love, has erected adequate obstacles to the mortal cold of European climates.

He was later to develop a more agressive style in his disputations.

---

[2]This publication, founded in 1789, contained numerous statistical essays.

Bienaymé's next paper in this area was the short *Naissances des enfants dans le mariage* (1837b), which was read to the *Société Philomatique* in December 1837. This paper concerns the statistical relationship between the number of children born in the first year of a marriage and the number later produced by those parents. From a mathematical point of view, it contains only some elementary arithmetic. It is also something of a sequel to the paper Bienaymé (1829), and the last paragraph reads as follows:

> M. Bienaymé then puts forward several reflections on the present imperfect state of statistics and on the inaccurate notions which former statistical works have propagated. One of these concerns the generally accepted view of the frightening mortality rate of infants. Undoubtedly newly-born infants die in greater proportions than children of 3 years or more. However, it was through taking no account of circumstances which should have played a part in the calculation, notably the continual increase in the population, that several earlier authors arrived at the extremely erroneous figure of 30 or even 40 deaths per 100 births in the first year.

No details on the reflections are given, a common feature of his *Société Philomatique* contributions, although apparently they follow something of the pattern of the 1829 note, being based on projection through to the mature age population. Certainly some of Châteauneuf's figures quoted in the 1829 note come within the range being criticized.

The principal results of Bienaymé's *Naissances* (1837b) paper are obtained from an examination of some of the records of marriages and of births in the period between 1809 and 1831. The conclusions are as follows:

> . . . the marriages of one year coincide, to the extent of at least half their number, with the births of the following year, and since the annual number of marriages is approximately one-quarter of the annual number of births, the marriages of one year produce at least one-eighth of the births of the following year. Again, the first year of a number of marriages taken at random produces more than one-eighth of the children which will result from these unions, although they may produce offspring for more than fifteen or twenty years.

Bienaymé claims that these indirect rules of thumb are preferable to direct investigation of a certain number of marriages and the children born in the marriages. This is because of difficulties in deciphering the state registers, and because a large number of families more or less promptly leave the community where the marriage has been contracted. Bienaymé's rules of thumb would additionally provide a check for gross error of a direct analysis.

The general use of such rules of thumb was not new. The difficulties of a direct census in the late 1700s and early 1800s had certainly been appreciated by Laplace, and earlier by Messance (1766) and Moheau (1778) (apparently pseudonyms assumed by de la Michodière and de Montyon, respectively). All regarded direct enumeration as an enterprise so difficult and expensive that it seemed unreasonable to try it, and proposed essentially the same indirect method. Laplace described the situation as follows in his *Essai philosophique sur les Probabilités*, which became the Introduction to his *Théorie Analytique des Probabilités* in the second edition (1814) [*Oeuvres* (1886), Vol. 7, Introduction, p. LI]:

> The registers of births, which are carefully kept to ensure records of the population, may be used to determine the population of a great empire without having recourse to counting the number of its inhabitants, which is a tedious operation and one which is difficult to perform with accuracy. For this purpose, however, it is necessary to know the ratio of the population to annual births.

In fact, Laplace proposed to the government that annual births and total population should be obtained for 30 administrative districts. His proposal was accepted and the population count was made on 23 September, 1802. Only those parishes whose mayors seemed sufficiently conscientious were used, and the number of births in each parish for the preceding 3 years was also recorded. From the resulting ratio and an estimate of the total number of births in France over an appropriate period, Laplace estimated the population to be [September, 1802] 28,352,485. He also claimed that the probability that this estimate is in error by more than 1/2 million was less than 1/300,000.

It is interesting to contrast Laplace's result with those of the censuses of 1801 and 1806, but first we shall give a little additional perspective.

One of the first acts of Lucien Bonaparte, who became Minister of the Interior in January 1800, was to prescribe a census—the results to reach him within 2 months. However, the difficulties of execution were greater than the will of the legislator, and it took not 2 months but 2 years for the results to come in. They were published in 1802 but do not appear to have been taken seriously. The census of 1806 seems to have suffered a similar fate. The official figures were, for 1801, 27,347,800; and for 1806, 29,107,-425. It is quite remarkable that Laplace makes no mention of these figures, which would have been well known to him, and this is an indictment indeed for the official results. It should be noted that Laplace had been Minister of the Interior before Lucien Bonaparte, albeit for a short tenure of only 6 weeks, and doubtless had an intimate knowledge of all the data of demographic statistics of his time. Further information on Laplace's work in demography can be obtained from Gillispie (1972), Stephan (1948), and Chang (1976).

Rather similar indirect methods were also used in Great Britain at about the same time. Sir Frederick Morton Eden (1800) estimated the population of Great Britain and Ireland via the number of births together with an estimate of the birth ratio, obtained with the aid of a sampling investigation to find the average number of inhabitants per house.

2.3. Life tables

Bienaymé's second paper was not read until February 1835, by which time he had risen in the Civil Service to the post of Inspector General of Finances. The paper, published as (1837a), is a major piece of work, discussing extensively and comprehensively the lifetables of Deparcieux and Duvillard, both of which were in wide use in France at that time.

In order to set Bienaymé's work in context, some historical comment on the demography of the period, in particular the work of Deparcieux and Duvillard, is necessary. For a more detailed account see Westergaard (1932).

Prior to 1800, the most accurate demographic material consisted of the records of religious orders and tontines, and it was on such records that the work of A. Deparcieux was based. He constructed the first life tables published in France. The subsequent ones of Dupré de Saint-Maur (published by Buffon in 1767) and Moheau (1778) were based on parish records, suffered from the problems of immigration, and did not achieve widespread use.

The tontine was a financial scheme whereby the subscribers to a fund each received an annuity which increased as the number of members was diminished by death, until the last survivor received the whole income. The idea, due to the Neapolitan banker Lorenzo Tonti, was proposed as a device for raising funds by the State. The members of the tontine would subscribe a sum needed by the State, and in return receive an annuity for life; the fund would return to the government after the death of the last surviving member. The idea was accepted into France with the tontine of 1653, and further state tontines were held in 1689 and 1696. The state drew little profit, however, as the probable duration of the annuities had been set too low. The government usage of tontines in France was finally banned by the king in 1763. Private tontines continued for some time, however, and there was a disasterous collapse of the tontine known as the *Caisse Lafage,* which had been established in 1791. Nevertheless, the demographic data that accrued through the tontine records was of particular value. See Hendricks (1863) for a historical account.

Deparcieux (1746) calculated mortality tables for the members of the tontines of 1689 and 1696, involving 5,911 and 3,345 persons, respectively, of whom 711 and 616, respectively, were alive in 1745. He also calculated a life table for the Benedictine congregation of St. Maur for the period 1607–1745. The monks concerned entered the order in the period 1607–1669, aged 17 to 25, and all of them died before 1745. It was therefore straightforward to calculate the numbers of persons exposed to death at each age, and hence the proper rates of mortality. Deparcieux, however, was additionally able to deal with extrapolation past the close of his period of observation, where the number of persons exposed to death at each age was not given.

The question of changes in the mortality rates did not appear to occur to demographers of this period.

Deparcieux was, not surprisingly, criticized on the question of the wider applicability of results based on the restricted classes of tontines and religious orders. At the time of his death, in 1768, he was engaged in the preparation of tables based on mortality records from 162 parishes in various parts of France, which had been collected for a period of 16 years. Subsequent attempts to treat this material were unsuccessful.

E. E. Duvillard was attaché of the *Contrôle Général* under Turgot, and later of the Ministry of Interior where, in 1805, he had charge of population statistics. This was at a time of considerable activity, but little real success, in population analysis. The French population censuses taken in 1801 and 1806 had been successful only in the provinces, and the results varied greatly. The estimates of experts furnished the largest portion of the material. An actual enumeration in Paris was not accomplished for the first time till 1817, following plans drawn up by Fourier. Duvillard, however, was motivated by the recent introduction of a scheme of vaccination against smallpox, and he took up the problems which D. Bernoulli (1766) had treated earlier. These concerned the lethality of the disease and finding the relationship between the number surviving who had never had smallpox and the number who had had an attack and recovered, on the basis of some life table. In his work (Duvillard, 1806) the mortality table was essentially no better than that of Bernoulli, who had effectively used Halley's (1693) table; it was constructed in 1798 on 101,542 deaths drawn from registers prior to the revolution and using the supposition of a stationary population. He was nevertheless of the opinion that his results were an accurate description of the situation in prerevolutionary France. It was not his principal aim to calculate a life table but rather to obtain means for comparison of the effects of vaccination. His conclusions were, for example, that, if smallpox disappeared, the average gain in life expectancy at birth would be 3 1/2 years.

Although Duvillard had restricted aims for his life table, it became widely used in spite of various objections. An important factor in this was that the indirect census taken in the year 1802 at the instigation of Laplace[3] gave a result for the ratio of annual births to total population very near to the result provided by Duvillard's tables (1/28.75). The methods used for this census, however, were not above serious criticism.

Bienaymé's object, in his paper *De la durée de la vie en France* (1837a), was to produce an overwhelming weight of evidence against the continued use of the Duvillard table. Although there had already appeared substantial criticism of the Duvillard table, for example, by Mathieu, in 1826, in *l'Annuaire du Bureau des longitudes,* many insurance companies still

---

[3]See §2.2.

based their calculations on the table, no doubt to considerable financial advantage, since the mortality rates that it predicted were much more rapid than was appropriate in France at that time. When Bienaymé wrote his article, the evidence available to him was that the average lifespan in France exceeded 36 years; the Duvillard table led to a result of 28.37 years. Bienaymé makes many comparisons of this kind in his paper. He clearly considered that he was doing an important public service in forcing the abandonment of the Duvillard table, as would seem an entirely appropriate action for an Inspector General of Finances.

It is interesting to note that the contrast that appeared between the life tables of Deparcieux and Duvillard was paralleled in Britain by the contrast between the 20 English Companies table and the Northampton table that it ultimately replaced. The former was quite similar to the Deparcieux table. and the latter to the Duvillard table. The following extracts (Bertrand. 1907. p. 310) give some comparison of these widely used tables. The starting point is a population of 890 at age 30.

Age	20 English Companies table	Deparcieux table	Duvillard table	Northampton table
30	890	890	890	890
35	854	841.50	820.59	813.89
40	813	796.63	750.30	737.78
45	769	754.20	678.54	659.23
50	718	704.48	603.38	579.87
55	657	637.79	522.39	496.86
60	584	561.40	433.78	413.64
65	491	478.95	337.93	331.24

The Northampton tables were produced by Richard Price (1771), the philosopher who communicated Thomas Bayes's famous essay (1763) on probability to the Royal Society 2½ years after Bayes's death. They were constructed from parish registers in Northampton and became a standard of British and American insurance companies until superseded by work of the Institute of Actuaries, nearly a century later, which produced the 20 English Companies table (Institute of Actuaries, 1869). Another table, based on population enumerations in Carlisle in 1780 and 1787 (Milne, 1815), was also rather widely used; it gave mortality rates closer to those of the Deparcieux table than those of the Duvillard one.

Bienaymé was aware of the work of Price and Milne and made some fourfold comparisons in his paper. For example, 60 individuals aged 20 are reduced to 50 by the age of 31 (Northampton table). 33 (Duvillard table), 40 (Carlisle table), 43 (Deparcieux table). The Northampton table, and similarly the Duvillard table, is statistically conservative for the insurer but, of course, some such effect is necessary to ensure the financial viability of insurance companies. One important ethical question concerns the dividing line between viability and profiteering.

It is interesting to compare the effect of the tables for annuities as opposed to life insurance. In this context the Duvillard and Northampton tables were conservative for the annuitant. In fact, in 1808, the British government, hard pressed by war and inflation, sought to raise finance by issuing annuities based on the Northampton tables. A substantial loss of about £2,000,000 in 11 years was suffered before the system was rectified, as the average lifespans of annuitants turned out to be considerably longer than those predicted by the tables.

Following this experience, a report was ordered by the House of Commons from John Finlaison and this appeared in 1824. He was then directed to make further investigations and *Report of John Finlaison* (Finlaison, 1829) resulted. This material embraced the results of various British tontines, these data not being subject to uncertainty over migration which the Northampton table (and the Duvillard table) would have suffered. It would have been of considerable interest to Bienaymé for comparative purposes, had he known of its existence.

On the other hand, there were cases in which the State, through the erudition of some of its bureaucrats, could benefit from the ignorance of its citizens. Thus, Hacking (1975, pp. 114–115) gives an interesting account of work of Hudde and de Witt in determining rates for the sale of annuities in the Netherlands. At the time, 1671, de Witt was chief statesman (grand pensionary) of the Netherlands, and Hudde was Mayor of Amsterdam. On 2 August, 1671, de Witt wrote to Hudde in the following terms:

> There is a general persuasion that the life annuity upon two lives, at 17 years' purchase, is much more advantageous than upon one life, at 14 years' purchase . . . it may even be that the joint annuity if sold at 18 years' purchase would be preferred to that upon a single life at 14 years' purchase; as this might produce a notable advantage to the republic, it is, in my opinion, of the highest importance to leave people in this persuasion; therefore I have not divulged it to anyone except yourself. . . .
>
> (cf. Hendricks, 1853, p. 101)

Doubtless Bienaymé would not have approved, nor indeed the people of the Netherlands. De Witt was not much longer to remain in power; he was forced to resign his post of grand pensionary on 4 August, 1672, and he and his brother were murdered by a mob of Orangists on 20 August, 1672.

Bienaymé did not achieve his aim, of discrediting the use of the Duvillard table in France at that time, unscathed. In the 1835 note he was forced to justify the accuracy of the data on which much of his 1837a paper was based. He says that he has written a second note which leaves no doubt about the small influence of errors in official documents on which his results were based. It is presumably this second note which appears appended to the 1837a paper as pp. 202–218. Furthermore, it appears that the use of the Duvillard tables continued for some time, and articles involving comparison of the Deparcieux and Duvillard tables, and criticism of the latter, continued to appear in the literature. In the forefront of this criticism was Demonferrand (1839a, b) and comparisons continued at least up to the work of Dupin (1849). Various modifications had been made to the tables by this time. For example, Demonferrand (1838, 1839a) had used the statistics of conscription for military service, a subject discussed in some detail in *De la durée de la vie en France* (Bienaymé, 1837a), as a means for correction of the tables.

The issue was one of scientific accuracy to Bienaymé, but it was economic necessity that preserved the use of that Duvillard table. The need for separate tables for companies for life insurance and for annuities was probably well appreciated by the 1840s, and the Duvillard and Deparcieux tables fitted these roles reasonably. Levasseur (1889, Vol. 2, p. 304) writes:

> In France, Deparcieux was used until 1888 to draw up tables for annuities and Duvillard to draw up life insurance tables. These tables, however, especially that of Duvillard, were no longer in accord with the current state of affairs, and actuaries were obliged to introduce corrections to them based on experience in order to establish premiums or compensation.

2.4. Probability and the law

The connection between probability and the law has a long history. It seems that the first major contributor was Leibniz (1665; revised, 1672). Leibniz had come from a legal family; his father and maternal grandfather were professors of law and he was trained as a lawyer prior to the development of his interests in mathematics. The issue in Leibniz's work is the connection between evidence and probability, and although he contributed little to the mathematics of probability, his conceptualization did have a lasting impact. Most of his contemporaries started with random phenomena, particularly gaming or mortality, and then speculated that the doctrine of chances could be applied to other cases of inference under uncertainty. Leibniz took numerical probabiity as a primary epistemic notion. Degrees of probability are degrees of certainty. For an account of this aspect of Leibniz's work see Hacking (1975, particularly Chapter 10). Hacking gives these ideas a highly significant role in the emergence of probability, and it is interesting to note that the decade around 1660 saw the real genesis of probability. In 1657 Huygens wrote the first probability textbook. At about this time Pascal made the first applications of probabilistic reasoning to problems other than games of chance,[4] and in so doing introduced the first ideas of a decision–theoretic kind, while Leibniz thought of applying metrical probability to legal problems. Graunt published (1662) the first

---

[4]See §5.8.

extensive set of statistical inferences drawn from mortality records. In the late 1660s annuities were put on a sound actuarial basis by Hudde and de Witt.[5]

Certainly Leibniz's work was noted by later masters; when Jacob Bernoulli considered the problem of evidence, he followed Leibnizian legal terminology, and Bernoulli's ideas on equipossibility (Part IV of his *Ars Conjectandi*, published posthumously in 1713; he died in 1705) appear to be derived from Leibniz.

The *Ars Conjectandi* was almost complete at the time of Jacob Bernoulli's death and was edited for publication by his nephew Nicholas Bernoulli, who himself wrote on probability applications to the legal system in a doctoral dissertation entitled *Specimina artis conjectandi ad quaestiones Juris applicatae,* published at Basle in 1709. This deals with many questions of a demographic kind and touches on issues of jurisprudence, such as the probability of innocence of an accused person for which Nicholas had a curious method of estimation. He assumed that any single piece of evidence given against an accused person was twice as likely to be false as true. He took different pieces of evidence as independent and thus computed the probability of innocence under n different pieces of evidence as $(2/3)^n$.

These were rather preliminary attempts to use the calculus of probabilities in the context of the legal system. The subject was first taken up on a major scale by Jean Antoine de Caritat, the Marquis de Condorcet. Condorcet was a marquis of the old nobility and a friend and disciple of Voltaire. His work was a product of its time. There was an increasing general concern with social justice, with the credibility of witnesses, with the question of whether a tribunal will reach a correct decision, and so on.

The relevant work of Condorcet (1785) is entitled *Essai sur l'application de l'analyse à la probabilité des décisions rendus à la pluralité des voix.* This work, which has recently been reprinted, consists of a *Discours Préliminaire,* which occupies CXCI pages, and of the five-part *Essai* itself, which occupies 304 pages. Condorcet's work is exceedingly difficult to follow. In Todhunter's (1865) words:

> The obscurity and self contradiction are without any parallel so far as our experience with mathematical work extends.

Nevertheless, Todhunter (Chapter XVII) gives a careful and detailed analysis of Condorcet's work; for further information see also Baker (1975), Gillispie (1972), and Maistrov (1967, 1974, Chapter 3, Section 8).

The *Essai* is largely concerned with voting systems and the closely related issues of decisions of juries and tribunals. The general conclusions that Condorcet draws from the first part of his work do not seem to be of

---

[5]For some comments on the work of Hudde and de Witt, see §2.3.

much significance; they amount to little more than obvious principles, though often based on unsound premises. For example, if the probability of a correct decision is the same for each voter and is greater than the probability of an incorrect decision, then the probability that the decision of the majority will be correct tends to 1 as the number of voters increases. This is an immediate consequence of J. Bernoulli's Law of Large Numbers, it being assumed that votes are independent.

Condorcet attaches great importance to his fifth question, which relates to the system of forced unanimity that operates for English juries. This question is discussed on pp. 267–276 and CXL–CLI. He believes that the system is bad and introduces the subject thus on p. CXL:

> Criminal trials in England are made in the following way: the jurors are obliged to remain on the court premises until such time as they are in agreement, and they are obliged to meet under this form of torture; for not only would hunger be a real torture, but boredom, stress and discomfort, carried up to a certain point, may become a veritable agony.

In the final part of Condorcet's book, he proposes what he thinks would be good forms of tribunal for the trial of civil cases and of criminal cases. For civil cases he suggests a court of 25 judges to decide by majority. For the trial of criminal cases, he suggests a court of 30 judges, in which a majority of at least 8 is required to condemn an accused person.

The next major figure in the area was Laplace. He came to Paris in 1768 and made the acquaintance of Condorcet shortly thereafter. Condorcet's ideals of social justice, as well as his mathematical applications, must have influenced Laplace. This was, of course, a time of social ferment and many members of the Academy of Sciences were active participants in the political maelstrom; the storming of the Bastille took place in 1789, and Condorcet met his death during the Revolution.[6]

A substantial portion of the second and later editions of Laplace's original (1812) treatise is relevant to the present section; the final version appears as Volume 7 of *Oeuvres complètes de Laplace* (1886), and it is to this that it is easiest to refer for present purposes (although we note that it was the basic 1812 edition which generally inspired Bienaymé). This portion comprises three sections in the Introduction,[7] titled *De la probabilité des témoignages, Des choix et des decisions des assemblées* and *De la probabilité des jugements des tribunaux*, the eleventh chapter entitled *De la probabilité des témoignages*, and two sections of the first supplement, *De la probabilité des jugements* and *Sur une disposition du Code d'instruction criminelle*.

The most significant of the work is in this supplement [1816] where Laplace is concerned with the probability that a juror will not be mistaken

---

[6]See §1.4 for references; especially valuable is David (1965).
[7]Which is Laplace (1814).

in arriving at a verdict of guilt; he assigns to this a uniform distribution on the interval [1/2, 1]. From this law he deduces the probability that the verdict is correct as a function of the majority obtained. A good discussion of the underlying mathematics is given in Meyer (1874, pp. 346–350).

Laplace's formula later became a source of argument when the jury system in France was in something of a state of flux. Stimulus for this had been provided by the commencement in 1827 of the publication of annual statistical reports by the Ministry of Justice.

Laplace's reasoning was sharply criticized by Poisson (1835), in which paper he laid down the principle that any chain of reasoning in this type of problem should be based on observations consequent on the operation of the Law of Large Numbers. The year 1835 also saw vigorous argument in the Chamber of Deputies on the jury law with the famous astronomer D.-F. Arago in the forefront. Arago, who had replaced Fourier in 1830 as permanent secretary of the mathematical sciences section of the Academy, had entered the Chamber of Deputies in 1831 as Deputy from Pyrénées-Orientales. In the argument, he drew heavily on the authority of Laplace and the name of science in claiming that a minimum majority of 7 against 5 for conviction would lead to a considerable number of errors. This was at a time when the juries of the Court of Assizes had 12 members. In the period from 1825 to 1830 a majority of at least 7 to 5 was required for conviction, but in 1831 it was changed to at least 8 to 4.

Poisson took up the issue again in another paper (1836). He attacked the problem from its foundations and examined the texts of laws as well as court records. He noted that in the period before 1831 the mean annual proportion of acquittals was 0.39 and that the proportion always lay between 0.38 and 0.40, while the mean proportion of convictions by a vote of 7 to 5 was 0.07. He then claimed that even before the required majority was changed in 1831 the effect of the new rules could be predicted. The proportion of convictions would be approximately $0.61 - 0.07 = 0.54$. This is indeed what happened in 1831.

Poisson's reasoning was treated with some scepticism and immediately after he had read his paper to the Academy of Sciences, Poinsot (1836) objected violently to the application of a calculus to the moral sciences. Poisson, however, continued unperturbed (1837a, b). Even in 1907, Bertrand was to write, rather unjustifiably, with reference to Poisson's reasoning,

What is one to make of all this? Absolutely nothing.

Nevertheless, in spite of the controversy, which is even now not entirely resolved, there was a surge of activity stimulated by Poisson, whose major work in the area is the famous *Recherches sur la Probabilité des Jugements* (1837b). This work, although designed as a contribution to juridicial theory, contains so much important preliminary material of a purely probabilistic nature that it must be regarded as a work on probability with applications based on the legal system. The German edition of 1841, which differs only slightly from the original, bears a much more accurate title: *Lehrbuch der Wahrscheinlichkeitsrechnung und deren wichtigste Anwendungen.*

Bienaymé entered the lists with a paper (1838a) read to the *Société Philomatique* in June 1838, which is alluded to by Todhunter (1865, §1043). Surprisingly, the paper itself does not explicitly mention Poisson, whose major work had appeared the preceding year. This is all the more noticeable since Bienaymé is concerned with commenting on the work of Laplace, whose conclusions had been rejected by Poisson. Bienaymé, intriguingly, does say in his last paragraph, however, that various (unspecified) reasons had prevented him from publishing his results earlier but that they had been communicated to various people, among whom was Liouville, who had mentioned them several times at a meeting of the Society. Liouville's journal in fact published a number of closely related papers in the same year as Bienaymé's paper (Cournot, 1838; Guibert, 1838).

In his paper, Bienaymé is concerned with demonstrating that the conclusions argued by Arago in 1835 were invalidly drawn from Laplace's work. He points to two kinds of errors in the formula of Laplace at issue. One of these had not been noted by Laplace; but the other had been commented on by Laplace, who took the view that he was making a sensible first approximation. The substantive error consists of attributing the same value to the probability that a judge is wrong when he pronounces the verdict guilty as when the verdict is not guilty. There is an inherent tendency to acquit the guilty rather than convict the innocent. Bienaymé endorses the ideas of Laplace but states:

> The theory would provide too large a number of unknowns for there to be enough equations capable of giving even approximate values.

The paper also contains a comment that is relevant to his criticism of Poisson's ideas on the Law of Large Numbers[8]:

> It is still understood here that all these variables are only arithmetical means of the opinions of all jurors. It can easily be shown that, as Jacob Bernoulli said, a multiplicity of causes produce the same effect as a single cause corresponding to the arithmetical mean of the possibilities which result from all these causes.

The only concrete juridicial conclusion in the paper is that the probability of error of a jury is larger for crimes against persons than for crimes where property is concerned. Nevertheless, while the tenor of the paper is negative, it represents a balanced criticism of earlier work quite unlike some of the diatribes that preceded it and were to follow.

---

[8]See §3.3.

Activities in this juridicial area were not confined to France. In Russia, M. V. Ostrogradsky presented a paper entitled "Extract from a memoir on the probability of judicial errors" to the Academy of Sciences of St. Petersburg in June 1834. It was published in 1838 and contained results in disagreement with those of Laplace. See Maistrov (1967; 1974, p. 181) for details. V. Ia. Buniakovksy's book whose title translates as *Foundations of the Mathematical Theory of Probabilities* (1846) had for its Chapter 11 the title "Applications of probability analysis to testimonies, legends, various kinds of choices between candidates and opinions and to court sentences by majority rulings" and contained some original work.

After the contributions of the late 1830s and early 1840s there appears to have been little activity in the area of juridicial theory and, indeed, comparatively little in probability as distinct from statistics. In connection with this period Gnedenko (1948, p. 394) writes:

> ... in spite of the fact that Laplace and Poisson concluded an important and fruitful initial period in the development of probability theory, a period of philosophical cementation of the basis of this science, this period resulted in an indifferent attitude towards probability theory in the West and in a definite rejection of the possibilities of utilizing its methods in studying natural phenomena. This led to the beginning of a long period of stagnation in probability theory in the West.

The following criticism by John Stuart Mill in his *A System of Logic* (1875; ninth edition, p. 67) seems to be representative of fairly widespread sentiment in the late nineteenth century. This appears in the context of applications of probability to the operation of juries and other tribunals.

> It is obvious, too, that even when the probabilities are derived from observation and experiment, a very slight improvement in the data, by better observations, or by taking into fuller consideration the special circumstances of the case, is of more use than the most elaborate application of the calculus of probabilities founded on the data in their previous state of inferiority. The neglect of this obvious reflection has given rise to misapplications of the calculus of probabilities which have made it the real opprobrium of mathematics.

Bertrand (1907) devotes the last chapter of his book to a discussion of the work of Condorcet, Laplace and Poisson. This is Chapter 13, entitled *Probabilités des Décisions*. He is quite scathing; surprisingly so, for one in the influential position of permanent secretary of the mathematical sciences section of the Academy of Sciences. He begins as follows:

> The boldest comparison, of sampling to the effects of unknown and variable causes, was proposed by Condorcet.
>
> The book on the probability of decisions made by a majority, which has been admired for too long, is based entirely on this confusion. None of these principles is acceptable. None of these conclusions approaches the true state of affairs.
>
> Condorcet's theory has been commented on, even entirely recast, by illustrious or famous scholars; no progress has been able to correct its inadequacy.

Condorcet's successors, while praising him for having brought the illumination of science to bear on these mysterious questions, have recognized the inadequacy of his formulae, but they have not put forward any better ones.

Laplace rejected the results of Condorcet; Poisson has not accepted those of Laplace. Neither has been able to subject to calculation a feature which is essentially absent: the chances of error on the part of a mind of greater or lesser enlightenment in the face of facts which are not completely known and laws which are imperfectly defined.

In the same vein, and somewhat more mildly, Markov (1924, p. 320) writes of:

. . . the inevitability of various arbitrary assumptions when dealing with this type of essentially undetermined problem. The problem becomes even more arbitrary if we allow that the witness may err and if we eliminate the assumption that the testimonies are independent.

See also the section entitled *Applications Erronées du Calcul aux Décisions Judiciaires* in Chapter 2 of Section E, *Les Probabilités* by E. Borel in *Encyclopédie Française,* Tome 1, Larousse, Paris (1937) (p. 1.92–14).

Against a background of all this criticism, it is interesting to note a recent resurgence of interest in the subject matter stimulated by the U.S. Supreme Court, and subsequently appearing in both the legal and statistical literature.

In 1970, the U.S. Supreme Court ruled that state juries need not be 12 in number and subsequently, in 1972, the need for unanimity in criminal trials was relaxed. From 1971 a six-man jury was introduced in civil trials in the U.S. District courts (see Pabst, 1974, for some comparisons with the traditional system). This changed situation promptly provoked papers in Law Review journals, for example, Zeisel (1971) and Walbert (1971), and in the statistical literature, where Gelfand and Solomon (1973, 1974, 1975) have reexamined Poisson's work and placed it in a modern setting. The statistical work has made extensive use of a large-scale study of judge and jury decision making by Kalven and Zeisel (1966).

2.5. Insurance and retirement funds

The first of Bienaymé's papers in this area is the paper (1839b) dealing with the effects of compound interest, read to the *Société Philomatique* in early 1839 and reported in purely verbal form.

The financial viability of such institutions as life insurance companies and retirement funds is affected essentially by two factors: (a) the force of compound interest; (b) random fluctuations in income and expenditure. Bienaymé claimed that the effect of compound interest in the consideration of stability of such institutions had hitherto been underrated.

Earlier, Laplace (1812, at the end of Chapter IX), and J. Bernoulli before him, had discussed the stability of insurance companies. Laplace had reached the conclusion, partly on the basis of increasing closeness of approximation of a probability by a relative frequency as the number of independent trials becomes large (Bernoulli's Law of Large Numbers), that for continued stability a very large amount of business is necessary, so that gains and losses essentially cancel each other out and fluctuation is confined to narrow limits. However, Bienaymé concludes that Laplace did not adequately consider the effect of compound interest, which results in the addition to the "large-numbers" requirement, the condition that it (the "large-numbers" requirement) be fulfilled in a relatively short time. This is necessary so that losses and gains in reserve funds cancel each other out in a relatively short time to insure that the effect of compound interest will not be significant. Bienaymé argues as follows:

> . . . a million francs payable in a hundred years would today be worth only about 7600 francs, the interest being 5%.

and then,

> . . . for a firm likely to gain or lose one million annually, the loss of a million from the first year could not be compensated by the gain of a million in the hundredth year, since this latter sum would be worth only about 7600 francs in cash. If there were no more recent income, an enormous reserve fund would be necessary.

Furthermore, the continued existence of an enterprise is all the less certain as the interest rate on which its calculations are based increases:

> Then, in fact, the number of years during which sums remain more or less comparable decreases considerably. If it requires 100 years at 5% for 1,000,000 francs to become reduced to 7600 francs, 51 years will be sufficient when the interest rises to 10%.

He then goes on to discuss some of the consequent social effects of compound interest in his usual penetrating and direct manner:

> . . . this increase in interest, which may provide a stimulus for the creation of insurance funds, pensions for widows and the aged and many other institutions worthy of the greatest consideration, becomes at the same time a source of advantage or a cause of ruin for these establishments. Continual vigilance is therefore needed on the part of administrators, and especially of enlightened minds among the founders, since the fate of large numbers of individuals depends on their foresight.

The importance of size of operation is noted:

> It is therefore very important to combine together in a single establishment as many operations as possible.

This is specially observed to be important in the case of retirement funds because of their limited amounts of interest per year.

Bienaymé, as always, seems to take particular note of the social implications of his results and he observes that the small businessman is most susceptible to the effects of compound interest, knowing that he cannot greatly enlarge his business. He is often accused of greed but cannot operate on the basis of Law of Large Numbers and has no alternative but to raise his profit margin.

Toward the end of his paper (p. 64), Bienaymé states that all his preceding discussion follows from a probability–theoretic formula of Laplace, *if one introduces into it the effect of compound interest*. Unfortunately the only direct clue to this formula is a reference by Todhunter (1865, §1041), in his discussion of Laplace's treatise (1812), to Cournot (1843, p. 333) but this contains no such formula. Cournot merely writes of:

. . . a grave error, which M. Bienaymé has rightly brought to light.

and proceeds to repeat some of Bienaymé's (1839b) comments, on the effect of compound interest, in his own words.

The formula of Laplace in question is presumably one on p. 431 of Laplace (1812). This refers to what appears to be the standard deviation whose size determines the length of a confidence interval for what is in fact the mean profit of the company per transaction. The issue is that, in the presence of compound interest, the divisor in Laplace's formula (viz. $\sqrt{s}$, where s is the number of transactions) should be replaced by the square root of the number of transactions in a single year multiplied by a quantity nearly inversely proportional to the interest rate (but rapidly approaching a limit as the number of years increases). The length of the confidence interval just becomes essentially a positive constant and may be reduced only by increasing the annual number of transactions, this being the variable divisor.

The paper evidences a respect for Poisson (mentioning *"sa pénétration habituelle"*) with whom Bienaymé had discussed the compound interest problem, and who had also reformulated it, in physical terms. This respect is also apparent in Bienaymé (1837b) where he writes that Poisson's *"discussion lumineuse . . . a bientôt levé cet obstacle"* but was to become less evident in Bienaymé's later writings, if unjustifiably so; a detailed discussion of the surrounding controversy is given in the next chapter.[9] It is possible (Stigler, 1974) that the relevant passage alludes to the Cauchy distribution which Poisson had devised in 1824. Indeed, the paper of Poisson (1824) is referred to by Bienaymé, in the context of an exception to the formula of Laplace, and the Cauchy distribution is introduced in that

---

[9]See §3.3.

paper by Poisson to provide an exception to Laplace's large-sample justification of least squares. Bienaymé's only direct reference to the Cauchy distribution occurs in his 1853c paper, following Cauchy's several papers where the density had been mentioned.[10]

Bienaymé's (and so Cournot's) view of compound interest, although more profound than that of Laplace, appears nevertheless to be simplistic. At the basis of the view is the belief that the institution in question, say an insurance company, keeps its incoming funds and reserves, as it were, "frozen." This view might have had some relevance at that time, but in more modern times the revenue would itself be invested in such a manner that it too earned compound interest at a rate at least that at which the insurance company credited its customers. Thus, the effect of compound interest either cancels out or works to the advantage of the company. An understanding of this principle is evident in Meyer (1874, p. 310).

Bienaymé did not return for many years (at least in his publications) to the subject of insurance companies and retirement funds. His last communications on the subject were a group (1857, 1862a, b, c, 1865) that all concerned the Metz mutual security society.

This series begins with a review (1857) of a monograph (report) of Colonel Isadore Didion (1855) on the *Société de Metz*. Colonel (General in 1858) Didion was a former professor of artillery at the *école de Metz* and a scientist of some note. Bienaymé draws attention to two facts in his review. The first is that in some 30 years of operation by the *Société de Metz,* the mortality rates experienced were in good agreement with those of the Deparcieux mortality tables. The second concerns the difficulty that the Society has experienced in coping with the illusion, produced by large accumulations of unencumbered capital, that all things were well concerning the safeguarding of its commitments and its continuity.

In the 1862a paper Bienaymé continues with his observations on the Didion report and remarks that the *Société de Metz* could easily find itself in serious financial difficulties:

> Proceeding from the basis contained in the [Didion] Report and the Deparcieux Table, it is found that the *Société de Metz,* despite the existence of its capital of 343,000 francs, has a deficit of about 100,000 fr. if the pensions of the present members are to be 200 fr., as are the 61 pensions already being paid. This deficit would be very much greater still if the subscriptions of its present members did not in future leave the same proportion available as in the past. As this is uncertain, it must be admitted that the actual deficit of the *Société* exceeds 100,000 fr. There is still time to remedy this unfortunate state of affairs, but one must cast aside all illusions and follow the advice of science and experience.

The points at issue here, together with the one of size of operation which is raised in the next quotation, were those highlighted in Bienaymé (1857).

---

[10]See §4.6.

The 1862a paper concludes with the following remarks:

> . . . it is to a large extent to protect Benevolent Societies from the dangers resulting from the promise of pensions by associations with too few members that the government has founded the Retirement Fund for old age, under the ministry of a Member of this Academy, M. Dumas, whose constant efforts in questions relating to the general welfare are known to all.

Bienaymé was himself much involved with the enterprise mentioned in the last paragraph and built up a substantial reputation for his efforts in this general area. As noted in §1.3, he was lavishly praised in a report read to the Senate on 26 April, 1864, by M. Dumas, then Minister of Commerce.

Bienaymé's remarks (1862a) were not long left unanswered, and officers of the Society responded in *Société de Metz* (1862) to which Bienaymé replied in two papers (1862b, c).

The main point made in the defense by the *Société de Metz* is that Bienaymé has not sought the reasons of the Administrative Council of the Society for not adopting the conclusions of the Didion report. They claim that Didion has underestimated the resources of the Society to the extent of neglecting interest of 8847 fr. 49 c. per annum on a capital of 190,351 fr. 34 c. Consequently, the capital is continuously increasing, and the Society is well able to cope with its expenses and pensions. They conclude by pointing out that they have also deposited successive sums now amounting to 29,000 fr. with the government retirement fund. The paper had the air of impassioned oratory and Bienaymé (1862b) responded briefly, immediately after the lecture. In this comment, Bienaymé admits that if there are errors in Didion's report then his own calculation would be affected but stands by the logic of his conclusions on the basis of the figures on which they have been constructed.

This exchange did not end the matter and Bienaymé again sought the permission of the Academy to speak about the *Société de Metz* and did so (1862c). He now takes a strong stance in reference to the claims of the Society:

> . . . it claims that its financial situation is very satisfactory in all respects. . . . Unfortunately this is by no means the case and I regret to say that, as I was certain in advance, the Report of Gen. Didion is completely accurate. Consequently, my calculations and my advice to a Society which is so concerned with its philanthropic objectives also stand in their entirety.

General Didion had asked Bienaymé to make it known that he did not make the errors asserted by the Society. Nevertheless Bienaymé manifests magnanimity in saying that his remarks concern not only the *Société de Metz*, but indeed the public good, for there are many such societies. In conclusion, he emphasizes that questions on assurance are scientific ones, since they can only be successfully based on the theory of probability, and claims that this was his basic reason for bringing the questions before the Academy. It is interesting that, in spite of his temperament, he rounds off the exchange on such a conciliatory note.

Bienaymé's connections with Didion were continued in some remarks (1865) in which he presents to the Academy a booklet on the calculation of pensions authored by Didion. This booklet, says Bienaymé, is a work which all societies concerned with pensions should consult. He does not, however, let the opportunity slip for some criticism of attitudes widespread in such societies:

> Unfortunately, the persons who direct these establishments which are so deserving of interest have occasionally neglected to too great an extent the suggestions of science, and this neglect has had grave consequences which it is important not to see repeated.

General Didion was a quite notable scientist and in the earlier exchange Bienaymé would undoubtedly have wished to clear him of responsibility for error and, indeed, to show that both of them were right. Bienaymé nevertheless appears through his work as a man of high moral principle.

3. Homogeneity and stability of statistical trials

> *History, regarded from the standpoint of unvarying causes, . . . offers most useful lessons to mankind.*
>
> P. S. de Laplace

3.1. Introduction

Homogeneity and stability, in their earlier historical setting, pertain in the main to independent binomial trials, where the probability of "success" in the ith trial is p_i. In the period of interest to us in this book, it was already well known that if $p_i = p = $ const., n trials are considered, and X is the number of successes, then as $n \to \infty$, $X/n \to p$ in probability, that is, $Pr(|X/n - p| \geq \epsilon) \to 0$ for any $\epsilon > 0$ (Bernoulli's Theorem). More recently, Poisson (1837b) had put forward the more general but heuristically motivated "rule" or "law," which he called the Law of Large Numbers, that even without the assumption $p_i = $ const., $X/n - \bar{p}(n) \to 0$ in probability, where $\bar{p}(n) = \sum_{i=1}^{n} p_i/n$. It was not rigorously demonstrated by Poisson and caused considerable confusion in regard to interpretation among the French probabilists, who therefore largely ignored it until the end of the century, although it had been proved (apparently unknown to them—surprisingly so in the instance of Bienaymé, in view of §1.4) rigorously by Chebyshev (1846). We shall generally refer to it as Poisson's Law of Large Numbers. Whereas Bernoulli's Theorem in an obvious sense expresses *stability* of X/n, under the assumption of constancy or *homogeneity* of the p_i, Poisson's law expresses, rather, the *loss of variability* of X/n.

The now-minor aspect of what is still often called the *dispersion theory of statistical trials* is concerned with considerations of the above kind which pertain to *stability of relative frequencies of statistical trials*. The second, and major, aspect consists of techniques developed for testing whether or not *homogeneity* of trials obtains and, if not, attempts to explain or typify the heterogeneity.

A large proportion of Bienaymé's papers deals with, or touches on, this area (Bienaymé, 1839a, b, 1840a, c, 1855). It seems most satisfactory, however, to incorporate them into what is a general, and not necessarily chronological, account of dispersion theory lasting well into the present century, in order to see and assess in context his contributions in his own time, and their eventual effect. Thus in this chapter also, he will tend to appear in a less major way, even though this is a major area of his activity.

3.2. Varieties of heterogeneity

The moment generating function, $M_X(t)$, for the distribution of the number of successes, X, in n independent trials, where p_i is the probability of success in the ith trial, was given in a modified form by Poisson himself:

$$M_X(t) = \prod_{i=1}^{n} (1 - p_i + p_i e^t).$$

From this it follows that the mean and variance of the distribution[1] are $EX = \sum_{i=1}^{n} p_i$, $\text{Var } X = \sum_{i=1}^{n} p_i(1 - p_i)$, generalizing the well-known Binomial distribution formulas for the homogeneous case (Bernoulli trials). Let us write:

$$\bar{p} \equiv \bar{p}(n) = \sum_{i=1}^{n} p_i/n, \qquad \sigma_p^2 \equiv \sigma_p^2(n) = \sum_{i=1}^{n} (p_i - \bar{p})^2/n.$$

Then:

$$E_2 X = n\bar{p}, \qquad \text{Var}_2 X = n\bar{p}(1 - \bar{p}) - n\sigma_p^2,$$

the last variance decomposition being well known (e.g., Aitken, 1957, pp. 50–52), and the subscript 2 having been introduced to distinguish from the simpler Bernoulli-case formulas when $p_i = p$: $E_1 X = np$, $\text{Var}_1 X = np(1 - p)$. Comparison of $\text{Var}_2 X$ with $\text{Var}_1 X$ shows that the variance in the Poissonian sequence can be said to be *less* than in the corresponding Bernoulli case.

Let us consider a variant of the above situation. Suppose one of a possible probabilities $p_1, p_2, \ldots, p_a$ is chosen, the choices being equiprobable, and then all n binomial trials carried out with the chosen probability of success. Then:

$$(3.1) \quad P(X = r) = a^{-1} \sum_{i=1}^{a} \binom{n}{r} p_i^r (1 - p_i)^{n-r}, \qquad r = 0, 1, 2, \ldots, n.$$

This distribution may be looked on as a posterior distribution (obtained from a uniform discrete prior distribution and Binomial conditional distri-

---

[1]It may be worth noting that other aspects of the distribution, such as the shape, are still objects of study (e.g., Darroch, 1964; Samuels, 1965).

bution) for the probability in a Binomial Distribution. In view of Bien-aymé's liking for Bayesian ideas, and his soon to be noted association with this variant, it is apt to do so. In relation to (3.1),

$$E_3 X = a^{-1} \sum_{i=1}^{a} np_i = n\bar{p}(a),$$
$$\mathrm{Var}_3 X = n\bar{p}(a)(1 - \bar{p}(a)) + n(n - 1)\sigma_p^2(a),$$

which formulas are deduced, e.g., by Aitken (1957, pp. 52–53). Note that $\mathrm{Var}_3 X$ *exceeds* the variance in comparable Bernoulli sampling.

Aitken mistakenly refers to the immediately preceding as the Lexian scheme. In fact, Lexian sampling, a later notion (Lexis, 1876, 1877, 1879), refers to m *sets* of n binomial trials, where the probability of success in the ith *set* is p_i, say, $i = 1, \ldots, m$. Another garbled account in respect to Lexian sampling occurs in Weatherburn (1947, pp. 115–116). In any case, to compare Lexian sampling with the preceding three schemes, we must consider for each of them m replications of n trials also; therefore, in each of the four schemes, we have a totality of $N = mn$ trials, in which we denote the number of successes by Y. The following table results:

Bernoulli: $E_1 Y = mnp$, $\mathrm{Var}_1 Y = mnp(1 - p)$
Poisson: $E_2 Y = mn\bar{p}(n)$. $\mathrm{Var}_2 Y = mn\bar{p}(n)(1 - \bar{p}(n)) - mn\sigma_p^2(n)$
Bienaymé: $E_3 Y = mn\bar{p}(a)$, $\mathrm{Var}_3 Y = mn\bar{p}(a)(1 - \bar{p}(a)) + mn(n - 1)\sigma_p^2(a)$
Lexis: $E_4 Y = mn\bar{p}(m)$. $\mathrm{Var}_4 Y = mn\bar{p}(m)(1 - \bar{p}(m)) - mn\sigma_p^2(m)$

Notice that the (true) Lexian scheme may also be viewed (by looking at the jth trial in each set, $j = 1, \ldots, n$) as n replications of a Poissonian scheme of m trials.

If we write P for the proportion of successes Y/N, then the $\alpha(t)$ level interval for P will be, for the ith scheme, $i = 1, 2, 3, 4$:

$$(3.2) \qquad E_i P \pm t \sqrt{\frac{\mathrm{Var}_i Y}{N^2}},$$

on assumption of asymptotic Normality of P. Evidently, for the scheme with subscript 3, the interval is:

$$(3.3) \qquad \bar{p} \pm t \sqrt{\frac{\bar{p}(1 - \bar{p})}{N} + \frac{(n - 1)}{N} \sigma_p^2},$$

which is longer than that for N comparable Bernoulli trials, differing by the second term in the radical, providing $n > 1$. Bienaymé (1839a) writes down the interval (3.3) and attaches to it (having, he says, obtained it "*par un calcul rigoreux*") the probability estimate:

$$(3.4) \quad \frac{1}{\sqrt{2\pi}} \int_{-t}^{t} e^{-x^2/2} \, dx + e^{-t^2/2} \bigg/ \sqrt{2\pi N\{\bar{p}(1 - \bar{p}) + (n - 1)\sigma_p^2\}},$$

analogously to Laplace's (1812, Chapter III) version of the Moivre–Bernoulli Theorem for Bernoulli trials, to which the expression reduces if $n = 1$. Note the first-order correction term; an elementary discussion of Laplace's (simpler) corresponding correction term is given by Uspensky (1937, especially p. 130). Bienaymé was led to his investigation by noticing that, indeed, in many experiments on repeated trials, P falls outside the interval given by the Moivre–Bernoulli–Laplace Theorem, when one would expect otherwise; so Laplace's hypothesis of constant probabilities seems rarely fulfilled in practice. Unfortunately, it appears that Bienaymé seemed never to conceive that deviation from the hypothesis of constant probabilities could also produce an interval (3.2) *shorter* than the comparable one for N Bernoulli trials, as with the Poissonian or Lexian schemes. Clearly, the phenomenon of P falling outside the Moivre–Bernoulli interval would in those cases occur only with negligible probability.

Bienaymé ascribes variation of probabilities to variation of causes and rationalizes the inhomogeneity of trials leading to (3.3) by his principle of *durée des causes:* suppose there are a causes, say $c_1, c_2, \ldots, c_a$, the ith cause giving rise to probability p_i of success, $i = 1, \ldots, a$. Each cause may occur equiprobably for any one of the m sets of n trials; but once chosen it *persists* for the whole set.

The case $n = 1$ is here particularly important, for although there are still a causes, there is no persistence of a cause beyond one trial, and the overall effect in the $N = mn = m$ trials is that of Bernoulli trials, with constant probability $\bar{p}(a) = \sum_{i=1}^{a} p_i/a$ of success in each. Bortkiewicz (1909) says that Poisson thought that the Bienaymé scheme, *even with $n > 1$*, was effectively one of such ordinary Bernoulli sampling, with $\bar{p}(a)$ as the effective success probability. Curiously, Bienaymé (1855) later made the same kind of error in regard to the Poissonian scheme, consequently claiming that Poisson's Law of Large Numbers was therefore just a trivial consequence of the Bernoulli Theorem.[2]

Bienaymé notes also that the additional term in the radical:

$$(3.5) \qquad\qquad (n - 1)\sigma_p^2/N$$

makes the interval (3.3) large or small depending directly on the size of n, that is, as *"la durée de la cause"* is large or small, respectively, and states:

> . . . the events about which we know most are precisely those which give us indubitable examples of this duration of causes.

He points out that if n is large (3.5) becomes $\approx \sigma_p^2/m$, so that the dominant term in (3.3) is of order $1/\sqrt{m}$ as n becomes large and hence, if m remains small, remains nonnegligible in comparison to the "Bernoulli term," which is of order $1/\sqrt{n}$.

---

[2]See §3.3.

We note, additionally, his remarks (Bienaymé, 1839a, p. 49) to the effect that he has worked and thought about such problems for 6 years; and that he has also considered the general case, where the a causes are not equiprobable and the lengths of the sets of binomial trials are not equal, but the formulas become complicated. Nevertheless, he says (of himself):

He will develop later the consequences of these formulas.

However, it appears he did not publish this further work. Indeed the first elements of such generalization are given by Bortkiewicz (1898)[3] with no mention of Bienaymé.

This paper (1839a) is especially notable for evidence of the influence of Laplace in Bienaymé's entire output on dispersion. Apart from apparently feeling the need to produce the correction term in (3.4), Bienaymé must needs mention on his p. 44 that Laplace (1812) himself, in his Chapter IX (under the title *Des bénéfices dépendant de la probabilité des événements futurs*) had modified the hypothesis of constant probabilities but had not pursued the topic in the direction of present interest.

It is clear from the above discussion that the third sampling scheme described above should bear Bienaymé's name, since the names mentioned for the other three schemes are now more or less accepted. Also, if the distribution (3.1) is to have a name at all, it should be Bienaymé's.

This paper (1839a) is closely associated with another (Bienaymé, 1855), and we shall take this up in the following section. It is, furthermore, his only paper to gain (somewhat belatedly) a place in the development of dispersion theory, at least among such effective heirs of Poisson and Bienaymé as Bortkiewicz and Chuprov. We take up amplification of this point now and simultaneously attempt to clarify a frequent notion propagated by these authors about the basic role of Cournot in the foundation of the theory. It shall be seen that this notion has come about largely through Bienaymé's association with Cournot and has little substance. We cite a number of passages:

Bortkiewicz (1909, p. 460):

. . . hypothesis of the duration of accidental causes, introduced by Bienaymé and Cournot.

Czuber (1910, p. 36, following a reference to an 1904 work of Bortkiewicz):

Bienaymé and Cournot were the first to cast doubt on the hitherto strongly held supposition.

---

[3]See §3.4.

Chuprov (1910, p. 280):

> . . . Bienaymé, with whom we associate also Cournot, appears indisputably as a distinguished predecessor of Lexis in the matter of the theoretical clarification of supernormal dispersion. . . . Bienaymé's investigations, however, do not go beyond the area of purely speculative assertions; he does not know how to reconcile theoretical considerations with concrete statistical material. . . .

Chuprov (1960, pp. 79, 227, within Russian translations of Chuprov, 1905, 1926):

> Using as basis the theory of . . . causes, developed by Bienaymé and Cournot, Bortkiewicz called this fact [stability and its behavior] "the law of small numbers" and gave it a precise mathematical foundation. . . .

> Bienaymé . . . developed . . . this conception [of Poisson and Cournot]. For statisticians . . . the meaning of this step remained incomprehensible. Even Bienaymé himself doesn't know how to make use of it, when he appears in the guise of a statistician . . . though as regards the theory, he reveals a very detailed grasp of the situation.

Chuprov (1918–1919, p. 199):

> . . . after Poisson, the conviction that $p \neq$ constant necessarily began to be accepted. This viewpoint . . . was caught up by Cournot and developed in the main by Bienaymé in forms serving as models for workers of future generations also. . . .

To what extent such comments eventually "evolved" is made clear by a portion of the introduction from a paper, close to the time of his death, by Bortkiewicz (1931, pp. 1–2):

> The idea of investigating the stability of statistical frequencies from the standpoint of the theory of probability goes back to the French mathematician Bienaymé. From various examples taken from social and moral statistics, he was the first to establish the fact that, almost without exception, the stability in question was essentially less than the "classical norm," that is, less than the expectation which is associated with the classical scheme of independent trials with a constant underlying probability. In order to explain this discrepancy between theory and observation, Bienaymé used a modification of the traditional procedure which was characterized by the assumption that between neighboring trials in a time ordered sequence a sort of dependence existed. Though interesting in itself and among other things adopted by Cournot as his own, we shall replace this method in what follows by another, originating from Lexis, which has the advantage of a wider usefulness, in that it can be applied not only to undulatory but to evolutory sequences.

The references to Cournot, which are given by Bortkiewicz, are to Cournot (1853, pp. 143 and 202).

In actual fact, Cournot merely *reports with due acknowledgement* in his §79 (pp. 143–145), albeit in a lucid manner, Bienaymé's (1839a) scheme of trials and corresponding formula (3.3), taking care to differentiate the essence of this scheme from schemes of other types. In §117 (pp. 202–205), where Cournot is concerned with general statistical aspects of sampling of this kind, Bienaymé is mentioned on p. 204 in the same connection; some of the sentences from this page are worth reproducing:

> In truth. within the province of statistical fact. the example of §79 is not directly applicable. One may not in general suppose that the chances remain strictly constant for an entire series of successive trials. and yet change sharply from one series to another.

In passing we should note that one of only two mentions of Bienaymé by Todhunter (1865, pp. 608–609, §1041) is in connection with expression (3.3) and cites Cournot (1843, p. 143).

There is further mention of Bienaymé by Cournot (1843) in a long footnote which begins on p. 201 and stretches to p. 203 and is connected with a final comment in his *Préface* (see our §1.3). This, however, refers to Bienaymé (1840c) and is not immediately relevant to the Bienaymé scheme, but to his method for testing for homogeneity of trials.[4] Bienaymé's name appears in this footnote on p. 202, and this is the point to which Bortkiewicz is inappropriately referring.

Some of the passages cited above are longer than necessary to make the point *re* Cournot, in order to reveal opinions that are otherwise relevant to this work. Overall, we see that, partly because of Bienaymé's laconic mathematics combined with rather overbearing verbosity, his ideas were misunderstood for a long time; when they were understood, moreover, they led to somewhat negative assessments of his statistical (as compared to probabilistic) ideas. We feel these are at least partly due to "benefit of hindsight"; the passages cited themselves reveal clearly the considerable likelihood of erroneous evolution of facts, through use exclusively of secondary and later sources, and one's own evolving memory of an early opinion.

3.3. Bienaymé and Poisson's Law of Large Numbers

Bienaymé's 1855 paper, the centerpiece of this section, subsumes his unpublished paper of 1842, which has already been mentioned in our §1.2 and whose title, only, is now known to us: *Communication sur ce qu'on doit entendre par la probabilité à posteriori*. On p. 199 [of the reprinted

---

[4]Taken up in §3.5.

version] he explains (rather in conflict with Bienaymé, 1852b, as is clear) that it was out of respect for Poisson (who died in 1840) that he had not published these findings at that time or earlier. He makes clear in no uncertain terms his continuing opinion that Poisson had not provided any relevant argument in support of his Law of Large Numbers, which does not exist, at least as a separate entity distinct from the Bernoulli Law. It is, he asserts, based on an illusion to which Laplace had not deigned to devote a chapter, and he, Bienaymé, accordingly, takes it upon himself to provide some *"explications indispensables."* He wishes his commentary to reach a wider audience than at the 1842 reading; hence his re-presentation of it. Later in the paper he is to deplore even the eyecatching name of the Poisson proposition, *Loi des grands nombres,* and to express the hope that the notion will not persist in the scientific literature.

Bienaymé thinks of binomial trials in connection with "causes." Furthermore, it is unrealistic to think of the cause, and so the probability of success, as not varying from trial to trial. However, if at each trial any one of a fixed number of causes can be operative equiprobably, and if each is associated with a fixed probability of success, then, in effect, the probability of success at each trial is the arithmetic mean of these probabilities, and so the relative frequency of success tends to approximate this arithmetic mean with increasing n, by Bernoulli's Theorem. In the notation of our preceding section, $X/n \to \bar{p}(a)$ in probability, and thus stability is manifested. Bienaymé's error is to think that the Poissonian concept amounts merely to this; in this he is unfortunately supported by the fact that the Poissonian intervals are no larger than those emanating from Bernoulli's rule. We have seen, in the previous section, the reason for this. He is also obsessed with the idea that any scheme that takes into account the variation of effective probability of success per trial must produce an interval longer than the Bernoullian; and yet, physical considerations lead him to believe that a fixed background set of causes should be associated with the trials. Both these requirements are fulfilled by his principle of *durée des causes,* which we have already discussed. In spite of high polemic and appeals to *"le bon sens"* against Poisson's Law, he does have the eventual modesty to admit that no principle is likely to explain all manner of variation of probabilities in trials.

If we write Poisson's Law in the manner of our §3.1:

$$(3.6) \qquad X/n - \bar{p}(n) \to 0 \quad \text{in probability, } n \to \infty,$$

we can see that Bienaymé is confused by the role of the arithmetic mean $\bar{p}(n)$ here, and the fact that this law does *not* in fact, as he thinks, express a tendency to stability of the frequency ratio X/n, in the general situation where a probability of success p_i is associated with the ith trial. In fact, no stability obtains in general (merely loss of variability), and his intuitive feeling that stability indeed cannot always obtain in such circumstances (see quotation below) lead him to question the law's whole foundation. The notion of a "floating mean," $\bar{p}(n)$, has ever since been an obstacle to intuition in applied work; and we shall mention a quite recent manifestation of this shortly.

On the other hand, there is no difficulty in the proof of (3.6) as a mathematical proposition. This was obtained by Chebyshev (1846); its proof is particularly easy by a direct application of the Bienaymé–Chebyshev Inequality, later used by Chebyshev (1867) to this end. It is hardly explicable that Bienaymé did not in 1855 know of the earlier paper of the two, since we have noted that his first ostensible contact with Chebyshev's work occurred in that year at the latest. Equally mysterious is Bertrand's (1889) ignorance: In his fine introduction, *Les Lois de hasard*, and also at the end of his exposition on Bernoulli's Theorem, he speaks of Poisson's Law as lacking in both precision and rigor and the presuppositions put forward by Poisson as unsuitable to serve as foundation for mathematical proofs by virtue of their lack of precision. This is in spite of the fact that one Frenchman (Laurent, 1873) had already given a proof of it, although Sleshinsky (1892), in his commentary on the history of the law, claims that Laurent's proof contains an error on p. 103, from line 13, which substantially alters the conclusion.

The law appears to have been a burning issue with Bienaymé. The report of Bienaymé (1845) shows, on its pp. 38–39, that he manages to give his cause support at this earlier time:

> Thus a population cannot continue in a stationary state as the first authors of mortality tables have supposed in their calculations. Precisely this, moreover, demonstrates the inevitable inaccuracy inherent in these calculations, and the pitfalls in almost all the consequently emergent ideas on the duration of life. Equally, it is a new example of a mean value to which large numbers of observations would not give constancy. In this connection, M. Bienaymé recalls that he has pointed out that M. Poisson had not demonstrated at all that which he thought to have proved under the name of the law governing *large numbers*.

The following passage from Quetelet (1848, p. 306) in relation to Quetelet's main text on p. 16 concerning the laws of society, evidences a continuing struggle by Bienaymé to understand the physical underpinnings of even the Bernoulli Law[5] and seems to reflect the Laplacian determinism as regards physical systems.

> A distinguished mathematician, M. Bienaymé, whose works have clarified several difficult points of the theory of probabilities, pointed out to me that my use of the term "law of accidental causes" was inappropriate, since what is being considered is events which definitely occur in a necessary and *a priori* calculable order, as the fluctuations which one notices have, really, nothing accidental about them if taken in sufficient number.

It is thus not altogether surprising to find a reprinting of the paper of Bienaymé (1855) in the organ of the Statistical Society of Paris in 1876,

---

[5]In this connection see also Bienaymé (1840b).

since at the time, Bienaymé was still alive and perhaps at the height of his influence.

Poisson's Law of Large Numbers finally takes its proper place in the theory and history of probability in the careful exposition of Czuber (1899), who also points out (perhaps for the first time?) the relationship between $\text{Var}_1\ X$ and $\text{Var}_2\ X$.

We conclude this section with mention of a return to all the hitherto discussed ideological difficulties of the law, within the context of a scathingly named and written article by Iastremsky (1957), who, himself, interprets the law correctly. While this article makes amusing reading, it is in the main an attack on the reactionary and bourgeois views of Lexis but has no mathematical foundation. Unfortunately, Iastremsky's mention of population genetics and of Schrödinger, in precisely the same critical context as Lysenko's (1951, pp. 20–21), characterizes it as belonging to the worst Soviet pseudoscientific writing of the period. Bienaymé is not mentioned, even though the other figures in dispersion theory receive some acknowledgement. It is evident why N. S. Chetverikov (1968) does not include Iastremsky's contribution in his collection, although he mentions it.

3.4. Dispersion theory

The preceding material suggests that, after Poisson, one should give a place to Bienaymé as a predecessor of the so-called "continental direction" of statistics, typified in particular by the later names of Dormoy, Lexis, his disciple Bortkiewicz, Chuprov, and Chuprov's student O. Anderson [1885–1960], to some extent Markov, and possibly including even Karl Pearson.[6] The main, and early, aspect of this "direction" is the study of the homogeneity and stability in repeated trials. A central idea is the construction of a test of whether the probability of an event in binomial trials remains constant.

To investigate whether this is so in a long series of binomial trials, Lexis (1876, 1877, 1879) suggested considering m disjoint sets (blocks) of trials, with n trials in each, where n may be small, although $N = mn$ is large. It is assumed that in the ith set the probability of success, p_i, is constant ($i = 1, \ldots, m$); the scheme has been partly discussed in §3.2. Denote now by P_i the actual proportion of successes in the ith set of trials, and as before, by P the proportion of successes in all N trials. Clearly,

$$P = \sum_{i=1}^{m} P_i/m.$$

---

[6] For the essence of this remark we are greateful to O. B. Sheynin (private communication, 1973).

The empirical criterion of investigation, or *statistic* as we would say in modern terminology, is the ratio:

$$D = \frac{\sum_{i=1}^{m} (P_i - P)^2/m}{P(1 - P)/n},$$

which we shall call the *empirical dispersion coefficient* following the general usage of the Lexis–Bortkiewicz school. Bortkiewicz tends to use $(m - 1)$ in place of m in its definition; this is virtually irrelevant in the sequel. A criterion involving a modulus, rather than a squaring, operation was proposed at approximately the same time by Dormoy (1878). Modern usage is to call D the Lexis quotient, or Lexis ratio.

To make simpler the consideration of the distributional properties of D, it was replaced by the slightly different variable:

$$D' = \frac{\sum_{i=1}^{m} (P_i - \bar{p})^2/m}{\bar{p}(1 - \bar{p})/n},$$

obtained from D by replacing P by $\bar{p}$ ($\equiv \sum_{i=1}^{m} p_i/m \equiv \bar{p}(m)$), this being thought allowable if N is large (and, as we now know, this is certainly so, by Poisson's Law of Large Numbers and Slutsky's Theorem, say). To find ED', for example, we see that:

$$E \sum_{i=1}^{m} (P_i - \bar{p})^2 = \sum_{i=1}^{m} E\{(P_i - p_i + p_i - \bar{p})^2\},$$

and ultimately,

$$= \frac{m}{n} \bar{p}(1 - \bar{p}) + \frac{n - 1}{n} m\sigma_p^2.$$

Thus,

$$ED' = 1 + \frac{(n - 1)\sigma_p^2}{\bar{p}(1 - \bar{p})}.$$

The quantity:

$$Q = \sqrt{ED'},$$

is called the *(theoretical) dispersion coefficient* or divergence coefficient (in German: *Fehlerrelation*).

Bortkiewicz (1909, p. 466) claims that he was the first to give this expression for Q in a rigorous manner in Bortkiewicz (1898), the prior demonstration of Lexis having been inadequate. The demonstration we have just given is, indeed, identical to that of Bortkiewicz (1898, Chapter 3, §14).

In the case of Bernoulli trials $p_i = \bar{p} = p = \text{const.}, \sigma_p^2 = 0$, and $Q = 1$, so with large N one may expect D to be close to unity. Substantial divergence of D from unity, then, suggests that $p_i \neq \text{const.}$ Note that, in any case, we are assuming *a priori* that the trials are independent and that there is a fixed probability of success in each block (p_i for the ith block); this has led to the conclusion that $Q \geq 1$—that the dispersion is *normale* ($Q = 1$) or *übernormale* ($Q > 1$).

In actual fact, there is nothing to prevent us considering the statistic D, and the variable D', in any m sets of n binomial trials, where the probabilities in individual trials can fluctuate in arbitrary manner. We may then, in general, define (with Lexis) a dispersion coefficient Q by:

$$Q = \sqrt{ED'}.$$

It is therefore of interest to us to calculate Q for the Poissonian and Bienaymé schemes of our §3.2. In the first case (for detail see Uspensky, 1937, p. 213),

$$ED' = ED_2' = 1 - \frac{\sigma_p^2}{\bar{p}(1 - \bar{p})},$$

where $\bar{p} = \bar{p}(n)$; so here $Q \leq 1$, and the dispersion may be *unternormale* ($Q < 1$). For the Bienaymé scheme, with the aim of computing $ED' \equiv ED_3'$, we consider the ith set of trials and denote by A_j the event that the probability of success in this set is p_j. Then,

$$\begin{aligned}
E\{(P_i - \bar{p})^2\} &= a^{-1}\sum_{j=1}^{a} E\{(P_i - \bar{p})^2 | A_j\} \\
&= a^{-1}\sum_{j=1}^{a} E\{(P_i - p_j + p_j - \bar{p})^2 | A_j\} \\
&= a^{-1}\sum_{j=1}^{a} \{E\{(P_i - p_j)^2 | A_j\} + (p_j - \bar{p})^2\} \\
&= (an)^{-1}\sum_{j=1}^{a} p_j(1 - p_j) + \sigma_p^2 \\
&= (an)^{-1}\{a\bar{p}(1 - \bar{p}) - a\sigma_p^2\} + \sigma_p^2 \\
&= n^{-1}\bar{p}(1 - \bar{p}) + \sigma_p^2(1 - n^{-1}),
\end{aligned}$$

where $\bar{p} = \bar{p}(a)$; so that,

$$E\left(\sum_{i=1}^{m} (P_i - \bar{p})^2\right) = mn^{-1}\bar{p}(1 - \bar{p}) + m\sigma_p^2(1 - n^{-1}),$$

and:

$$ED' \equiv ED_3' = 1 + \frac{(n - 1)\sigma_p^2}{\bar{p}(1 - \bar{p})}.$$

Thus, for the Bienaymé scheme,

$$Q \equiv Q_3 = \sqrt{1 + \frac{(n-1)\sigma_p^2}{\bar{p}(1-\bar{p})}},$$

which is the same expression as for the Lexian scheme, except that here $\bar{p} = \bar{p}(a)$ and in the Lexian scheme $\bar{p} = \bar{p}(m)$. If in fact $a = m$, it follows that Q is the same for both schemes.

The identity of Q for the Lexis and Bienaymé schemes was pointed out by Bortkiewicz (1898), in his *Anlage* [Appendix] 2, pp. 42–48, where he carries out an analysis of the above kind for the Bienaymé scheme, and in fact permitting a general distribution for the choice of success probability out of a possible choices for each set of m trials, having already considered the Lexis scheme in his §14. However, Bienaymé is not mentioned in this book at all, nor is *durée des causes*. Indeed, Bortkiewicz regards what we have described as the Bienaymé scheme as one of m sets of "dependent trials," in the physical sense that the choice of a p_i for one set fixes the success probability for all n trials of the set. We find, however, in a footnote on p. 467 of Bortkiewicz (1909) the following:

For comparison of results furnished by this method and that of I. J. Bienaymé see *L. von Bortkiewicz* Gesetz der Kleinen Zahlen p. 42/8 (Anlage 2).

As an aside, we mention that the book of Bortkiewicz (1898), *Gesetz der kleinen Zahlen*, has secured for itself a permanent niche in the history of probability, in connection with its study of the Poisson distribution and the famous Prussian cavalry "deaths-from-horsekicks" data. The title of the book, *The Law of Small Numbers*, while evidently in contrast to Poisson's Law of Large Numbers, has caused considerable difficulty in the interpretation of what it actually describes. We agree with Mises–Geiringer (1964) in thinking it refers to the Poisson probability distribution ("the Law of Rare Events"). Bortkiewicz does not seem to define the term in the book itself. An interpretation by Chuprov has been mentioned in our §3.2.

Before leaving Q to return to the basic statistic D and its properties, we mention that in a study involving *dependent* trials in each of m sets, but the sets being independent of each other, Chuprov (1905) showed that both the cases $Q > 1$ and $Q < 1$ could arise from this situation. One of his schemes is in fact the Bienaymé one with $a = 2$ choices for p_i, which can take on the values $p_1 = 1$, $p_2 = 0$.

With the work of Chuprov and Markov in dispersion theory in the period 1913–1922, attention reverted to a consideration of the statistic D itself and its probability distribution under the *a priori* (null) hypothesis that the $N = mn$ trials are all Bernoulli (i.e., independent, with a fixed probability of success in each). Markov, in the 1913 edition of his book *Ischislenie Veroiatnostei*, proposes the replacement of D by the modified statistic:

$$L = \frac{N-1}{n(m-1)} D,$$

where, recall, $N = mn$. (The multiplier thus introduced is required only to avoid bias, in the modern manner.) L is defined as 1 if $P = 0$ or 1. Later Markov (1916) shows that:

$$EL = 1$$

(this even if the number of trials in each of the m sets is not the same). He credits this result to Chuprov, without a specific reference, which is, however, essentially to Chuprov (1916), a publication more or less simultaneous with Markov's (on the basis of an earlier private communication), but whose proof is more complex—see Chuprov (1918–1919, §I). Markov alludes also to Bortkiewicz's work pertaining to ED' and derives, additionally, an expression for Var $L \equiv E\{(L-1)^2\}$, obtaining finally the simplification:

$$\text{Var } L \leqslant 2/(m-1), \quad \text{if } m \geqslant 5.$$

This material is reproduced by Markov (1924) and Uspensky (1937). Markov's paper of 1916 itself clarifies the interaction between himself, Bortkiewicz, and Chuprov, and we reproduce, at the cost of some repetition, the relevant passages here:

> Without turning to the question of the role of the dispersion coefficient in statistics, I intend in the present note to firstly prove the assertion of Professor Chuprov, that the expectation of the coefficient of dispersion in the case of independent trials with constant probability is precisely unity, if one does not take square roots and confines oneself to the definition in my book *Ischislenie Veroiatnostei* (1913); and, secondly, to determine for the case of repeated series a sufficiently simple approximate expression, actually, an overestimate, for the mathematical expectation of the square of the deviation of this coefficient from unity.
>
> I associate the assertion about the expectation of the dispersion coefficient with the name of Professor Chuprov because, as far as I know, he was the first to investigate the fraction as a whole and not the numerator and denominator separately; and arrived at the above-mentioned conclusion, at least for the case of identical series.
>
> In regard to the second question, a kind of solution to it was found long ago, by Professor L. Bortkiewicz, but this is associated with assumptions unacceptable to us, quite apart from problems of precision, clarity of definitions and full rigor of derivation.

In a sequel, Markov (1920) goes on to derive the asymptotic distribution of L, as $n \to \infty$ (and even in the more general case where the sample sizes may not be the same for each set, but are each made to approach infinity). This paper, which seems to be unknown—even Uspensky (1937, p. 219) in giving an account related to ours is unaware of its existence—asserts that:

$$\lim_{n \to \infty} P[L \le l] = \{\Gamma(\alpha)\beta^\alpha\}^{-1} \int_0^l y^{\alpha-1} e^{-y/\beta} \, dy,$$

with $\alpha = (m - 1)/2 = \beta^{-1}$. Note that this is a very simple form of the Gamma Distribution and that the simple linear transformation of L to:

$$(m - 1)L$$

will render the asymptotic distribution of the new variable to be $\chi^2(m - 1)$ —Chi-square with $m - 1$ degrees of freedom. A historically interesting derivation of the same result based on Moivre's Normal approximation to the Binomial is reported in Mises (1964, p. 440).

In the general tenor of this section, let us now examine the statistic L from a more modern viewpoint. Suppose we have m sets of numbers arranged into groups of n. Write x_{ij}, $i = 1, \ldots, m$, $j = 1, \ldots, n$, for the jth number in the ith group, and put also $N = mn$,

$$\bar{x}_i = \sum_{j=1}^{n} x_{ij}/n , \qquad \bar{x} = \sum_i \sum_j x_{ij}/N.$$

The sum-of-squares decomposition,

$$\sum_{i=1}^{m} \sum_{j=1}^{n} (x_{ij} - \bar{x})^2 = \sum_{i=1}^{m} \sum_{j=1}^{n} (x_{ij} - \bar{x}_i)^2 + n \sum_{i=1}^{m} (\bar{x}_i - \bar{x})^2,$$

is well known from simple one-way analysis-of-variance theory for testing homogeneity of m independent samples of n. Returning to our scheme of m sets of n trials, if we write X_{ij} for the random variable which takes the value 1 if the jth trial in the ith set (group) results in success, and 0 otherwise, we find since $X_{ij} = X_{ij}^2$ that:

$$NP(1 - P) = n \sum_{i=1}^{m} P_i(1 - P_i) + n \sum_{i=1}^{m} (P_i - P)^2.$$

Thus,

(3.7) $$L = \frac{S_b^2}{m - 1} \bigg/ \frac{S^2}{N - 1},$$

where

$$S_b^2 = n \sum_{i=1}^{m} (P_i - P)^2, \qquad S^2 = NP(1 - P),$$

i.e.,

$$L = \frac{N - 1}{m - 1} \left\{1 + \frac{S_w^2}{S_b^2}\right\}^{-1},$$

where

$$S_w^2 = S^2 - S_b^2 = n\sum_{i=1}^{m} P_i(1 - P_i).$$

Thus, the asymptotic distribution of the statistic:

(3.8) $$Y(m,n) = \frac{S_b^2}{m-1} \bigg/ \frac{S_w^2}{m(n-1)},$$

as $n \to \infty$, follows easily from Markov's result, since the asymptotic distribution of $(m - 1)Y(m,n)$ is clearly that of $(m - 1)L$, which we have seen to be $\chi^2(m - 1)$ distributed, under our *a priori* assumption that all $N = mn$ trials are Bernoulli. Thus, either of the statistics:

$$S_b^2 \bigg/ \frac{S^2}{N-1} \quad \text{or} \quad S_b^2 \bigg/ \frac{S_w^2}{m(n-1)}$$

may be used as a test criterion for the homogeneity of m series, each of n independent trials, providing n is large, by use of the tabulated $\chi^2(m - 1)$ distribution.

A number of important points may consequently be made, concerning the historical role of the analysis of this section. First of all, we may consider D in the light of the following decomposition:

$$mD = \frac{\sum_{i=1}^{m} (nP_i - nP)^2}{nP(1 - P)}$$
$$= \sum_{i=1}^{m} \left\{ \frac{(nP_i - nP)^2}{nP} + \frac{(nQ_i - nQ)^2}{nQ} \right\},$$

where $Q_i = 1 - P_i$, $Q = 1 - P$. If we regard our m sets of n trials each as samples from m populations of values in which elements take only the values 0 and 1 (e.g., when we are investigating absence or presence of a certain attribute in each individual), then we recognize the last expression as the *Chi-square statistic* for testing the *homogeneity* of these populations; and we now know from general theory that, as $n \to \infty$, under the null hypothesis of homogeneity, the asymptotic distribution of mD is the $\chi^2(m - 1)$ distribution. Although this result accords with those of Markov just described, the first important point is that Lexis's ratio is essentially a case of, and anticipates, the various Chi-square statistics. R. A. Fisher (1928) was probably the first to notice this in 1924. Further, in his book (R. A. Fisher, 1958, p. 80) there is a statement to the effect that the measure of dispersion of Lexis is χ^2/r, the number r being (p. 76):

. . . the number of classes the frequencies of which may be filled up arbitrarily, without altering the expectations.

K. Pearson (1900) was the first to obtain the Chi-square distribution as the asymptotic distribution of a Chi-square statistic. This was in connection with the "goodness-of-fit" criterion for a frequency curve; the Chi-square statistic and its asymptotic distribution for testing the homogeneity of several populations came later (Lancaster, 1969). (Markov's contribution in respect of this last, has, as noted, apparently been hitherto unnoticed.) It is also worth noting that in 1930 S. Kolodziejczyk showed that the criterion based on mD for testing such homogeneity followed from the likelihood criterion, whence the asymptotic $\chi^2(m - 1)$ distribution. (In fact he examined the same general situation as considered by Markov in his derivation of the asymptotic distribution, namely, unequal class sizes all made to approach infinity.)

Of course, the continuous $\chi^2(r)$ distribution for the sum of squares of r independent standardized Normal variables was known much earlier. Sheynin (1966) gives the credit for its discovery to E. Abbé (1863), although Lancaster (1966, 1969) points out a strong case for assigning the discovery to Bienaymé (1852a).[7] The concept was popularized some years later by Helmert (to whom until recently the discovery had been attributed). The opinion of Seal (1967) is that Lexis proposed his coefficient after Helmert and was not aware of any connection between the two ideas. This, he says, did not come until Bortkiewicz (1922) had:

> . . . given his qualified approval of the use of χ^2_{m-1} for the approximate distribution of the Lexis coefficient.

There is, in our opinion, very little concrete evidence for this assertion in Bortkiewicz's paper, in comparison to Fisher's claim to the insight.

The interested reader should consult Lancaster (1966, 1969) for detailed and extensive information about the development of Chi-square statistics and the related role of the Chi-square distribution (concerning whose historical role see also Sheynin, 1971).

The second important point concerning Lexis's ratio is the fact that the statistic $Y(m,n)$ defined by (3.8) has the form of an F statistic in single-factor analysis-of-variance. The assumptions in the present context are not, of course, the usual normality assumptions of that theory, in which $Y(m,n)$ would have an $F(m - 1, m(n - 1))$ distribution under the null hypothesis that all mn readings come from the same Normal distribution, whether n is large or not. Nevertheless, it is well known that if an F statistic with an $F(m - 1, m(n - 1))$ distribution is multiplied by $m - 1$, and then n is allowed to approach infinity, an asymptotic $\chi^2(m - 1)$ distribution results; and we have seen this to be the case with $(m - 1)Y(m,n)$ as $n \to \infty$. The generalization of Lexis's ratio (3.7) to arbitrary measurements, i.e., where:

$$S^2 = \sum_{i=1}^{m} \sum_{j=1}^{n} (X_{ij} - \bar{X})^2, \qquad S_b^2 = n \sum_{i=1}^{m} (\bar{X}_i - \bar{X})^2,$$

---

[7]See §4.3.

is, indeed, generally attributed to Bortkiewicz (Bortkiewicz, 1909?). It is evident, then, that the usage of the Lexis–Bortkiewicz quotient as a test criterion directly anticipates one-factor analysis-of-variance. In *that* theory, the quotient arises as a test statistic directly from the likelihood criterion, under the usual independence and normality assumptions on the X_{ij} (which in particular are assumed to have common variance σ^2). It is these distributional assumptions, implying as they do independent Chi-square Distributions for S_b^2/σ^2 and S_w^2/σ^2 under the null hypothesis, that lead to consideration of the modified statistic (3.8) in place of the quotient (3.7). Even in the generalized case, and under the assumption that the X_{ij} form an independent sample of $N = mn$ from an *unspecified* distribution with well-defined finite variance σ^2, the fact that $S_b^2/(m - 1)$, $S^2/(N - 1)$, and $S_w^2/m(n - 1)$ each have expectation σ^2 seems to have been understood from the 1920s. Eventually, Geiringer (1942b) even obtained the generalization to this setting of Chuprov's result, that $EL = 1$; and also that $EV = 1$, where:

$$V = \frac{S_w^2}{m(n - 1)} \Big/ \frac{S^2}{N - 1}.$$

There exists a number of accounts of the influence of Lexis's ratio on more recently developed areas of statistical theory. Most are listed in the fine sketch of Lexis by Heiss in the *International Encyclopedia of the Social Sciences*. A particularly interesting and more extensive account is that of R. K. Bauer (1955), who was apparently greatly influenced by Oskar Anderson, whom we have already mentioned as a figure in the continental direction. Although Bauer does not mention either Poisson or Bienaymé (whose scheme of binomial trials is nevertheless discussed as "Model B"), he is generally informative and we summarize his main points. The influence of the Lexis ratio is traced through to the theory of analysis-of-variance and experimental design, C. R. Rao's dispersion analysis, and multivariate analysis. These latter areas can be described as being of the English statistical school, founded by K. Pearson. The approach of this school may be summarized by the statement that precise prior distributional assumptions, often of normality, apart from some parameters, are made in respect to the data, and their consequences developed. It is acknowledged that attempts to mitigate these assumptions have been made within the school, through the study of "robustness," generalized curves, and transformations; to this we add that for large sample sizes, through the existence of appropriate limit laws (especially for the likelihood ratio), the role of prior distributional assumptions is minor. On the other hand, according to O. Anderson and Bauer, in the continental direction of consequences of Lexis's theory (whose main proponent was Chuprov), it is sought to give test criteria of more universal applicability by avoiding precise distributional assumptions. In our view, the main area of present day statistical theory which embraces this continental view is that of nonparametric statistics; and the contributions of more recent Russian researchers in the area, such as Kolmogorov and Smirnov, are well known. To these names one should add that of Mises. Other manifestations of the continental tradition, according to Bauer, are the variate-difference method and discriminant analysis.

While the above distinction is possibly artificial in view of the role of limit theorems ("large-sample theory") as a unifying element of all mathematical statistics, it may well be said that from a viewpoint of applied fields (and Oskar Anderson, as well as Lexis and Bortkiewicz, may be regarded, like Cournot, as economists or econometricians), the continental view may be the more sensible for moderate samples, refined statistical theory being unjustified.

It is not surprising, in view of her connection with Mises, that the Anderson viewpoint is particularly exemplified in several of the writings of H. Geiringer (e.g., 1942b). She explicitly seeks to steer a middle course between highly specific distributional assumptions and asymptotic distributional results applicable only to large samples, by attempting to present an "exact" analysis-of-variance exposition mentioned above. An extensive understanding of Oskar Anderson's views and methodological directions may best be obtained directly from his own textbook (Anderson, 1954), which is otherwise notable for its historical references. Indeed, Bienaymé is mentioned 11 times, but, regrettably, only in connection with the (Bienaymé–Chebyshev) Inequality; and Cournot is mentioned eight times, but not in connection with dispersion theory.

For the reader interested in the evolution of Lexis's (or dispersion) theory, the best source is N. S. Chetverikov (1968) which contains in Russian translation by the editor, from the original German, the (possibly) key papers of Lexis (1879), Bortkiewicz (1894–1896), Chuprov (1918–1919), and Bauer (1955), together with a commentary by the editor on Lexis's paper. Useful secondary accounts are given also by Czuber (1910), Polya (1919), Rietz (1932), Uspensky (1937), Geiringer (1942a), Bernstein (1946a), Gini (1955, 1956), and Mises (1964), some of which differ from ours in detail and all of which omit Bienaymé in the context of dispersion theory.

For biographical sketches of Bortkiewicz and O. Anderson, see accounts by Sheynin in the *Dictionary of Scientific Biography*. For Chuprov, see Kohn's account in the *Encyclopedia of the Social Sciences*. An excellent obituary of Anderson is given by Wold (1961).

We thus conclude our sketch of *dispersion theory,* apart from a neglected contribution of Bienaymé.

3.5. Bienaymé's test

Much of the preceding section is concerned with testing the null hypothesis of constancy of probability in a long sequence of binomial trials. Bienaymé (1840d) addresses himself to the same problem.

He first shows that if we take, as given, a successes in c trials, then split the c trials into successive blocks of n_1 and n_2 $(n_1 + n_2 = c)$, then (under the null hypothesis) the distribution of the number of successes r in the first n_1 trials is given by the hypergeometric formula:

$$(3.9) \qquad \binom{n_1}{r}\binom{n_2}{a-r} \bigg/ \binom{c}{a}, \qquad r = 0, 1, \ldots, \min(a, n_1),$$

which is independent of the postulated common success probability, p. Later he shows this to remain true if the n_1 trials for consideration are chosen randomly (but without replacement) from the totality of c trials. The concordance of the obtained value of r with the null hypothesis is then to be tested using a version of the Central Limit Theorem in the form:

$$(3.10) \qquad P\left\{-t < (r/n_1 - \bar{p})/(n_1^{-1}\bar{p}(1 - \bar{p})(1 - n_1/c)) < t\right\}$$
$$\approx (1/\sqrt{2\pi}) \int_{-t}^{t} e^{-x^2/2}\, dx + e^{-t^2/2} \bigg/ \sqrt{2\pi\bar{p}(1 - \bar{p})(n_2/c)n_1}$$

with $\bar{p} = a/c$, providing c is large and n_1 is large $(n_1 \leq a)$.

Although we shall return to the point soon, we mention now that it is evident that Bienaymé in 1840 may be aware of the entire theory, finite and asymptotic, of random sampling without replacement in a population of individuals where each is considered from the viewpoint of possessing a certain attribute or not ("success" or "failure"). He calls the totality of the above, including the independence of p, a *principe entièrement nouveau* in one of the titles of his paper.

Bortkiewicz (1918, especially p. 116), writing partly on the theory of drawing balls without replacement and, thus, on the theory of the Hypergeometric distributions, erroneously thinks Bienaymé is concerned with the estimation of $\bar{p} \equiv a/c$, the actual proportion of "success" balls in the totality of c, and correctly asserts that it is not at all surprising that Bienaymé's formulas for drawing inferences on $\bar{p}$ do not involve $\bar{p}$, but rather the relative frequency r/n_1, which is what Bienaymé thought to be quite new in Bortkiewicz's opinion. He makes reference to Bienaymé's mere statement of the various expressions and criticizes him for giving the impression that (3.10) follows in quite similar manner to the Laplace formula for Bernoulli trials. He concludes that Bienaymé's idea of investigating homogeneity in a sequence of trials on the basis of a scheme corresponding to ball drawing without replacement is put in too abrupt a contrast to other methods with the same aim, and that its practical significance is overemphasized by him; nevertheless, he admits the idea to be important from a purely theoretical viewpoint.

At this stage we are in a position to conclude, perhaps somewhat tentatively, that Bortkiewicz spent a great deal of time attempting to understand of what, precisely, Bienaymé's (1839a, 1840d) laconically put mathematical expressions were a manifestation. It is from Bortkiewicz, moreover, that Chuprov's comments and opinions *re* Bienaymé are evidently derived; so it is clearly Bortkiewicz's long-term difficulties that are responsible for Bienaymé's small place in dispersion theory, although it may be argued that without Bortkiewicz, Bienaymé's work in the area may not have been noted at all.

We have already noted Cournot's (1843) mention of this paper of Bienaymé at the end of his *Préface*. The point is taken up again in Cournot's footnote stretching over pp. 201–203, in each case the *L'Institut* version of Bienaymé's paper being cited. Cournot notes, with Bienaymé, that the distribution of r/n_1, given $\bar{p} = a/c$, will be independent of p and therefore bear no relation to it; r/n_1 will not, thus, be close to p in general even if large numbers are involved. His emphasis is not on the hypergeometric mathematics, nor on the testing for constancy of p, as is Bienaymé's; but on the fact that *given a series*, one cannot use a *partial series* to draw inferences about it. That is, in observing independence from p, given $\bar{p} = a/c$, in (3.9), both Bienaymé and Cournot are approaching the concept of a *sufficient statistic*[8] for p.

We return to the implications for hypergeometric sampling of the various expressions of Bienaymé. The precise expression for the variance of the distribution (3.9) is now well known and can be stated in modified form as:

$$\text{Var}(r/n_1) = \frac{\bar{p}(1 - \bar{p})}{n_1} \left(1 - \frac{n_1}{c}\right) \bigg/ \left(1 - \frac{1}{c}\right),$$

with $\bar{p} = a/c$, whereas Bienaymé gives the expression:

$$\frac{\bar{p}(1 - \bar{p})}{n_1} \left(1 - \frac{n_1}{c}\right),$$

which is also cited at the end of Cournot's *Préface*. Again, Bienaymé (but not Cournot) has in place of $r/n_1 - \bar{p}$ [in the expression corresponding to (3.10)] the quantity $r/(n_1 + 1) - (a + 1)/(c + 2)$. From the asymptotic nature of (3.10) these differences may be dismissed. Second, the Central Limit Theorem for sampling without replacement from a finite set of numbers does not seem to have an extensive history; it may be traced from the article of Bikelis (1969), who himself is concerned with obtaining a *uniform* bound on the remainder term, whereas Bienaymé's remainder term in (3.10) is of more sensitive kind, depending, as it does, on t. In particular, the Central Limit result without remainder specifically for the *Hypergeometric distribution*, which we are considering, is given in the various editions of the textbook of Bernstein (e.g., 1946a, p. 256). In any case, we can see that (3.10) is in itself a substantial achievement for its time.

In conclusion to this section, it is interesting to note some later, and apparently independent, work related to Bienaymé's, and to the whole problem of homogeneity of binomial trials, by Campbell (1859). The pre-

---

[8]See §5.6.

sentation of this is within a socially motivated framework, but its essence is as follows. Suppose we are *given:* (i) m (>1) replications of n Bernoulli trials, so that the total number is $N = mn$; and (ii) the fact that the total number of successes is $a = mb$, where b is an integer. Then the (conditional) probability of obtaining r successes exactly in any one specified replication of n trials is:

$$\binom{a}{r}\binom{N-a}{n-r}\bigg/\binom{N}{n}, \qquad r = 0, 1, \ldots, \min(a,n),$$

which can be written in alternate form as:

$$\binom{n}{r}\binom{N-n}{a-r}\bigg/\binom{N}{a}, \qquad r = 0, 1, \ldots, \min(a,n),$$

in agreement with Bienaymé's (3.9). Note that the expected value of this distribution is b; Campbell deduces that b is also the most probable (or, as we would now say, modal) value of the distribution. Thus, in a list of N trials split into blocks of n, the number of successes in successive blocks is, likely, b or near integer neighbors of b, so that if the trials are Bernoulli, the m blocks will resemble each other in this respect; this conclusion is the purpose of Campbell's paper. He notes also that if, in a given block, an "extreme" value of r occurs, one can (by the now well-known calculation of tail probability using the Hypergeometric Distribution) thus test the hypothesis of overall homogeneity of trials. We note two further interesting features of Campbell's analysis aimed at avoiding awkward hypergeometric situations; he lets $n \to \infty$, and $m \to \infty$ keeping a/m constant, respectively, and obtains (in effect) the respective limit distributions:

$$\left(\frac{m-1}{m}\right)^a \binom{a}{r} (m-1)^{-r}, \qquad r = 0, 1, \ldots, a;$$

and

$$\binom{n}{r}(b/n)^r (1-(b/n))^{n-r}, \qquad r = 0, \ldots, n,$$

the second of which, he notes, might have been foreseen. Note that both distributions are Binomial, the first with parameter $1/m$.

4. Linear least squares

Truth is the daughter of time.

R. Bacon

4.1. Introduction

In order that Bienaymé's papers on this subject appear in their proper perspective, we shall give a brief history of the earlier probabilistic aspects of linear least squares, particularly as developed by Gauss and Laplace, in the next section. Both these authors are frequently mentioned in Bienaymé's contributions, which are, naturally, strongly motivated and colored by the work of Laplace. The work of Gauss in this area is well described by Plackett (1949) and Seal (1967). Before proceeding, a few remarks at this stage about the general nature of the motivating problem will help elucidate the whole development of this chapter. We shall use modern matrix notation in our exposition and shall otherwise simplify the original mathematical notation where convenient. It is well to note, however, that use of modern notation obscures many of the difficulties encountered by the original workers.

The basic problem of interest is to estimate an $r \times 1$ vector, $\boldsymbol{\beta}$, of unknowns from a number N of observations $\mathbf{Y}$, related linearly to $\boldsymbol{\beta}$ but subject to error $\boldsymbol{\epsilon}$:

$$(4.1) \qquad \mathbf{Y} = X\boldsymbol{\beta} + \boldsymbol{\epsilon}.$$

Here $X = \{x_{ij}\}$ is a known fixed $N \times r$ matrix with $N \geq r$, which is assumed to be of full column rank r. The manner in which the mathematicians of the nineteenth century regarded the problem of estimation was to find an $r \times N$ matrix $K = \{k_{ij}\}$ (or, as they would have it, a system of "multipliers" k_{ij}) such that:

$$(4.2) \qquad KX = I,$$

where I is the unit matrix. Then $\boldsymbol{\beta}$ was, accordingly, estimated by:

$$\tilde{\boldsymbol{\beta}} = K\mathbf{Y} = \boldsymbol{\beta} + K\boldsymbol{\epsilon},$$

the matrix K being chosen [under the constraint (4.2)] in some optimal manner.

The problem may be regarded still, as it essentially was then, in two ways. The first viewpoint is nonstatistical and regards the problem as one of *interpolation*, (4.1) being considered as an overdetermined set of linear equations. In this case K is determined in accordance with a *direct requirement on ϵ itself*. The second viewpoint is *probabilistic*, K being chosen in accordance with some *distributional requirement on ϵ*. (It will be convenient, for example, to speak later of the Bienaymé–Cauchy controversy first from one viewpoint and then from the other.) The "least-squares" choice of K is:

$$(4.3) \qquad\qquad K = (X'X)^{-1}X',$$

with corresponding "least-squares" estimate of β:

$$\hat{\beta} = (X'X)^{-1}X'\mathbf{Y}.$$

As is well known and shall be reiterated in the next section, this choice may be justified from both viewpoints, interpolational and probabilistic.

4.2. Legendre, Gauss, and Laplace

In 1805 Legendre deduced the choice (4.3) on purely interpolational grounds, that is: K in (4.1) is to be chosen subject to (4.2) and in such a way that the sum of squares of errors $\epsilon'\epsilon$ is minimized. While giving no probabilistic justification, he did, however, pertinently stress that the usual confident summarization of the information in n observations $x_1, x_2, \ldots, x_n$ by a single number, their arithmetic mean, $\bar{x} = \sum_{i=1}^{n} x_i/n$, *follows from the same Method of Least Squares*, since it is readily seen that the value of μ which minimizes the sum of squares:

$$\sum_{i=1}^{n} (x_i - \mu)^2,$$

is indeed $\mu = \bar{x}$. There is a kind of reciprocal to this point, made by Gauss (1809) and sometimes called the *principle of the arithmetic mean*. This states that if for a sequence of n (independent) observations $x_1, \ldots, x_n$ each described by a probability density f satisfying certain presupposed conditions, the arithmetic mean $\bar{x}$ is the most probable combination of $x_1, x_2, \ldots, x_n$ for all possible numerical values of each of $x_1, x_2, \ldots, x_n$ and each $n \geq 1$, then:

$$f(x) = (2\pi\sigma^2)^{-1/2} \exp(-x^2/2\sigma^2), \qquad -\infty < x < \infty,$$

for some positive number σ^2. That is, the common distribution is the Normal, with zero mean.

In his first (1809) justification of the least-squares choice of K, Gauss uses this principle of the arithmetic mean as the basis for assuming that the errors ϵ_i are i.i.d. (independently and identically distributed) and ϵ_i is $\mathfrak{N}(0,\sigma^2)$. Many authors have commented on the artificiality of this justification for the assumption of normality, and Gauss himself was not unaware of this. A progressively more acceptable justification for the normal as the "Law of Error" is its role as the limit law of an increasingly large (standardized) sum of "elementary errors," i.e., the Central Limit Theorem, generalized to arbitrary i.i.d. observations by Laplace (1812, Chapter IV). In any case, with these assumptions, ϵ is $\mathfrak{N}(0,\sigma^2 I)$, so that its probability density is:

$$f(\epsilon) = (2\pi)^{-N/2} \exp\{-\epsilon'\epsilon/2\sigma^2\}, \qquad\qquad -\infty < \epsilon < \infty,$$
$$= (2\pi)^{-N/2} \exp\{-(Y - X\beta)'(Y - X\beta)/2\sigma^2\},$$

from (4.1). Thus, if one chooses that estimate for β which *maximizes* this density (i.e., the "most probable posterior" estimate; or, in modern terms, the "maximum likelihood" estimate) then one clearly arrives at the least-squares estimate $\hat{\beta}$. (Gauss also obtained the appropriate maximum likelihood, $\equiv$ the "generalized least-squares," estimate of β, allowing that the ϵ_i may not have identical variances, but we shall not linger on this heteroscedastic case here or later.) It is worth mentioning in passing that, according to Sheynin (see Maistrov, 1967, 1974; §3.10), a principle of the arithmetic mean as well as a theoretical foundation for the Method of Least Squares were obtained slightly before Gauss by the now little-known American mathematician Adrain (1808).

In his second justification Gauss (1821) continues to assume that the ϵ_i are still i.i.d., but with unspecified common density $\phi(\epsilon)$ satisfying $\phi(\epsilon) = \phi(-\epsilon)$, and, implicitly, having well-defined variance σ^2. He seeks a matrix K satisfying (4.2) and such that the mean-square error:

$$\int_{-\infty}^{\infty} \cdots \int_{-\infty}^{\infty} \left(\sum_{j=1}^{N} k_{ij}\epsilon_j\right)^2 \phi(\epsilon_1)\phi(\epsilon_2) \ldots \phi(\epsilon_N) \, d\epsilon_1 d\epsilon_2 \ldots d\epsilon_N,$$

is minimized for arbitrary i, $i = 1, \ldots, r$. Note that the expression may be written:

$$(4.4) \qquad \int_{-\infty}^{\infty} \cdots \int_{-\infty}^{\infty} (\bar{\beta}_i - \beta_i)^2 \phi(\epsilon_1)\phi(\epsilon_2) \ldots \phi(\epsilon_N) \, d\epsilon_1 d\epsilon_2 \ldots d\epsilon_N,$$

where $\{\bar{\beta}_i\} = \bar{\beta}$; so we actually seek a $\bar{\beta}_i$ which minimizes the expression subject to (4.2). Gauss shows that this will be so only if $\bar{\beta}_i = \hat{\beta}_i$, $i = 1, \ldots, r$, where $\hat{\beta}$ is the least-squares estimate. Thus we can take $K = (X'X)^{-1}X'$. It is clear that in (4.4) we have just an expression for Var $\bar{\beta}_i$, since $E\bar{\beta}_i = \beta_i$ [in view of (4.2)], so that we are examining, among the class of unbiased linear estimates of β_i, those which have minimal variance irrespective of the true value of β_i; and it turns out that the least-squares estimate fulfils these conditions for all $i = 1, \ldots, r$. The generalization of this result to arbitrary linear estimable functions of β even when X is not of full rank is now very well known as the Gauss–Markov Theorem (for details, see Seal, 1967). We pause to note, for future use, that the covariance matrix of $\bar{\beta}$ is, in general,

$$E((\bar{\beta} - \beta)(\bar{\beta} - \beta)') = E(K\epsilon\epsilon' K')$$
$$= \sigma^2 KK',$$

so, in view of (4.3), in the least-squares case:

$$E((\hat{\beta} - \beta)(\hat{\beta} - \beta)') = \sigma^2 (X'X)^{-1}.$$

Thus, the variance expressed by (4.4) is the ith diagonal element,

$$\sigma^2 \sum_{h=1}^{N} k_{ih}^2, \qquad i = 1, \ldots, r.$$

Temporally intermediate to the two justifications of Gauss are two of Laplace (1812). In the first of these, the framework is much like that of Gauss immediately preceding—unspecified $\phi(\epsilon)$, which is symmetric and has finite variance; actually the density is assumed confined to a finite interval $(-g, g)$. An extensive discussion, along the lines of Laplace himself, is given by Meyer (1874, Appendix I, pp. 355–376). Laplace shows (at least, for $r = 1$) that the standardized random variable:

$$\sum_{h=1}^{N} k_{ih}\epsilon_h \bigg/ \left\{ \sigma^2 \sum_{h=1}^{N} k_{ih}^2 \right\}^{1/2},$$

has approximately as $N \to \infty$ an $\mathfrak{N}(0,1)$ distribution. It follows that, for a fixed probability level, a symmetric interval about the origin for the error $\sum_{h=1}^{N} k_{ih}\epsilon_h \equiv \bar{\beta}_i - \beta_i$ is then of the form:

$$\pm t \sqrt{\sigma^2 \sum_{h=1}^{N} k_{ih}^2},$$

so that the confidence interval for β_i engendered by it will be shortest when:

$$\sum_{h=1}^{N} k_{ih}^2,$$

is minimal, which again leads to the Method of Least Squares.

Laplace's second justification, like Gauss's later, has a decision–theoretic flavor about it, except that he chooses to minimize:

$$(4.5) \quad \int_{-\infty}^{\infty} \cdots \int_{-\infty}^{\infty} \left| \sum_{j=1}^{N} k_{ij}\epsilon_j \right| \phi(\epsilon_1)\phi(\epsilon_2) \ldots \phi(\epsilon_N) \, d\epsilon_1 d\epsilon_2 \ldots d\epsilon_N$$
$$= E\{|\bar{\beta}_i - \beta_i|\},$$

of course subject to (4.2), i.e., $\sum_j k_{ij}x_{jr} = \delta_{ir}$ (Kronecker's delta). He uses the approximate $\mathfrak{N}(0,1)$ distribution for standardized error when N is large, deduced in his first justification, to note that then, in fact, (4.5) is:

$$2\sigma^2 \sum_{h=1}^{N} k_{ih}^2 \Big/ \sqrt{2\pi\sigma^2 \sum_{h=1}^{N} k_{ih}^2} = \sqrt{(2/\pi)\sigma^2 \sum_{h=1}^{N} k_{ih}^2} ,$$

so, again, the requirement of minimization of the length of a (one-sided) interval for the error leads to least squares.

We thus see that while Gauss's methods justify least squares for any fixed N, Laplace's are asymptotic as $N \to \infty$ and tend to avoid distributional problems regarding $\bar{\beta}_i - \beta_i$ in the same way as the Central Limit Theorem does.

4.3. Bienaymé's contribution

In the opinion of Bienaymé (1852a), the large-sample theory is the only one of practical applicability, and hence his preference is in the direction of Laplace. Thus also, while in the main Laplace's results were considered for $r = 1$, with some discussion of $r = 2$, Bienaymé gives essentially our above description of Laplace's first justification for general r (in this connection see also Todhunter, 1869). Again, a faithful and detailed account of this work is given by Meyer (1874, Appendix II, pp. 377–408). This brings the degree of generality in line with Gauss's exposition for arbitrary r. In an attempt to manifest the innate strength of the Laplacian approach, Bienaymé forcefully makes the valid point that the asymptotic normal theory of that approach is valid irrespective of the chosen K, so long as (4.2) holds.

However the real purpose of Bienaymé's paper arises from his observation that, in dealing with multiple linear regression ($r \geq 2$), one may easily obtain a confidence interval, at a given level of probability, for any one coefficient β_i. However, what had not been considered to date, and thus provides a *"lacune à combler"* (lacuna to fill), is the important and more relevant problem of *finding a simultaneous confidence region for all coefficients* β_i, $i = 1, \ldots, r$. This is certainly the relevant problem if one is not just interested in one specific coefficient, since composition of univariate confidence intervals is invalid (as they are generally not independent and in any case the type I error probability is magnified unacceptably). During his introductory §1, he indicates that the tools for tackling the problem have already been provided by Laplace (first justification), and that the philosophical approach of Laplace is the only "haven from all objections." His dislike of the first justification of Gauss comes about from its normality assumption, and of the second because Gauss based the use of the Method of Least Squares only on "considerations" while "real" proofs are to be found only in Laplace. It is relevant for us to remark that while Gauss's theory has become firmly entrenched in "classical" statistics, Laplace's is almost forgotten, and possibly rightly so. In this instance, Bienaymé's championing of Laplace's cause is not really consistent with the realities even of his own time. [That loyalty is an issue may be evidenced by a reference to Cournot (1843) for a table of values of the normal integral, rather than to some other source.]

Section 1 of Bienaymé's paper is particularly notable, as an indicator to his personality and state of excitement, for the following passage in relation to his proposed idea of the simultaneous confidence region:

> There will be more than one obscurity to be reexamined within the applications of the method of least squares. However here we shall consider only one which is most frequently evident, and which we now specify in manner so simple that within the first words the whole world will recognize the existence of the problem, even though the modifications it necessitates lead to quite complicated analysis.

In a book published well after Bienaymé's death, Bertrand (1889, §226) takes strong exception, and not without some reason, to this statement and several related ones. He says in part:

> Bienaymé . . . has proposed "a profound modification" . . . he speaks of the "defect of ordinary calculation." The flaw which he announces seems to him so simple "that within the first words the whole world will recognize the existence of the problem."—"The error consists," he says, "in calculating the probability of an error committed as if it were the only one . . ."

While conceding the importance of Bienaymé's problem, Bertrand says it is wrong to accuse the authors of the theory of least squares and its applications of having ignored or forgotten it. Indeed, if one is only interested in one coefficient, Bienaymé's idea is irrelevant: it all depends on the problem at hand.[1] We add that, indeed, there is evidence emanating from §4 of Bienaymé's paper that Bienaymé himself thought, incorrectly, that his considerations were of relevance even for a single coefficient.

The bulk of Bienaymé's actual calculations is contained in §2 of his paper. These are well described in §4 of Lancaster (1966), from which the following account differs only in some minor details. Let us write for fixed but arbitrary K satisfying (4.2):

$$\tau = K\epsilon = \tilde{\beta} - \beta.$$

---

[1] One of the examples that Bertrand uses to illustrate his point is related to a note (Bertrand, 1888b) in which he (almost) proves that $E(S^2 \mid \bar{X} = \bar{x}) = ES^2$ for a normal sample. This is a forerunner of the famous result on statistical independence of S^2 and $\bar{X}$.

Bienaymé forms the joint characteristic function $P(\alpha)$ of the error vector τ (a device used by Fourier and Dirichlet before him):

$$(4.6) \quad P(\alpha) = \int_{-\infty}^{\infty} \ldots \int_{-\infty}^{\infty} \exp\left(i\alpha'\tau\right)\phi(\epsilon_1)\phi(\epsilon_2)\ldots\phi(\epsilon_N)\, d\epsilon_1 \ldots d\epsilon_N,$$

and observes that the joint density of the error vector τ is given by the inversion:

$$Q(\tau) = (2\pi)^{-r} \int \exp\left(-i\tau'\alpha\right) P(\alpha)\, d\alpha,$$

so that the probability p corresponding to a specified subregion. R, of the sample space of τ is given by:

$$(4.7) \qquad\qquad p = \int_R Q(\tau)\, d\tau.$$

In theory, then, the problem of a simultaneous confidence interval for the β_i, $i = 1, \ldots, r$, in terms of $\bar{\beta}_i$, $i = 1, \ldots, r$, is resolved, since $\tau = \bar{\beta} - \beta$, if one can evaluate such integrals, at least as $N \to \infty$. However, even then, since in practice one *specifies* a p and is concerned with finding an *appropriate* R, the problem is not well defined; and what Bienaymé resolves ultimately to do is to deal only with R of certain (tractablé) form, once given p.

In regard to (4.6), he notes that he has to deal with a product of N factors, each of the form:

$$\int \exp\left(is_h\epsilon_h\right)\phi(\epsilon_h)\, d\epsilon_h,$$

where, since $\tau = K\epsilon$, $s_h = \sum_j \alpha_j k_{jh}$, and writing $\mu_j = E(\epsilon_h^j)$, the integral becomes:

$$1 + i\mu_1 s_h - \mu_2 s_h^2 - i\mu_3 \frac{s_h^3}{4} \ldots ;$$

or, in exponential form, using cumulant notation,

$$\int \exp\left(is_h\epsilon_h\right)\phi(\epsilon_h)\, d\epsilon_h = \exp\left\{ iK_1 s_h - \frac{K_2 s_h^2}{2!} - i\frac{K_3 s_h^3}{3!} + \frac{K_4 s_h^4}{4!} \ldots \right\},$$

so that:

$$P(\alpha) = \exp\left\{ iK_1 \sum s_h - \frac{K_2}{2!} \sum s_h^2 - i\frac{K_3}{3!} \sum s_h^3 + \frac{K_4}{4!} \sum s_h^4 \ldots \right\}.$$

To proceed to (4.7), Bienaymé makes a transformation from τ to ρ where

$$\rho_i = \left(\tau_i - \mu \sum_h k_{ih}\right)/\sqrt{2\sigma^2}$$

(where $\mu = \mu_1 = E(\epsilon)$—he gives various reasons for not assuming $\mu = 0$, the usual supposition—$\sigma^2 = \text{Var } \epsilon = \mu_2 - \mu_1^2$), and a number of further linear transformations of variables to arrive at the following version of (4.7):

$$p = \pi^{-r/2} \int dt \, \exp(-\mathbf{t}'\mathbf{t})[1 - \frac{1}{6} B_3(\mathbf{t}) + \frac{1}{24} B_4(\mathbf{t}) \ldots],$$

where the $\mathbf{t} = \{t_i\}$, $i = 1, \ldots, r$, are related linearly to τ and are to be integrated over a region corresponding to R. The $B_i(\mathbf{t})$ naturally depend on the form of $\phi(\epsilon)$, the initial error density.

At this stage Bienaymé *decides to concentrate on simultaneous confidence regions R tantamount to the region:*

$$\mathbf{t}'\mathbf{t} \leqslant \gamma^2,$$

which, he says, is one of the tractable situations. He gives a reason for the $B_3(\mathbf{t})$ playing no part in the last expression for p; and for neglecting further B_i if N/r is large, following a device of Laplace. Assuming this to be so, he obtains:

$$p = \pi^{-r/2} \int dt \, \exp(-\mathbf{t}'\mathbf{t})$$

integration being over $\mathbf{t}'\mathbf{t} \leqslant \gamma^2$, for fixed positive γ. It is important for us to note that this integral is effectively one whose integrand is the joint density of r i.i.d. $\mathfrak{N}(0,\tfrac{1}{2})$ r.v.'s, irrespective of how Bienaymé arrived at it. With the aid of the theory of the beta function he expresses p as:

$$p = \frac{2}{\Gamma(r/2)} \int_0^\gamma u^{r-1} e^{-u^2} \, du,$$

where $u^2 = \mathbf{t}'\mathbf{t}$.

Thus, if Z_i, $i = 1, \ldots, r$, are i.i.d. $\mathfrak{N}(0,1)$ random variables, and we write:

$$U^2 = 2^{-1} \sum_{i=1}^r Z_i^2,$$

it follows by Bienaymé's result that:

$$p = Pr\,[U^2 \leqslant \gamma^2],$$

and hence:

$$Pr\left[\sum_{i=1}^r Z_i^2 \leqslant x\right] = \frac{1}{2^{r/2}\Gamma(r/2)} \int_0^x y^{(r/2)-1} e^{-y/2} \, dy,$$

which is the distribution function of the Chi-square distribution with r degrees of freedom [$\chi^2(r)$ distribution]. From this, Bienaymé's claim to a place in the discovery of this distribution is self-evident.[2] Further, Bien-

---

[2]See §3.4.

aymé gives expressions for evaluating p when $r = 2n - 1$ (with the aid of the normal integral), and $r = 2n$, and later in the paper gives some values. Lancaster (1966) has pointed out that these expressions are essentially the same as K. Pearson used to construct his tables.

Bienaymé's transformations are such that for a fixed p (or, equivalently, γ) the greatest and smallest values the transformed error ρ_i may take, corresponding to the confidence region, are (Bienaymé, §3):

$$(4.8) \qquad\qquad \pm\gamma\sqrt{\sum_h k_{ih}^2},$$

(so this interval is rendered shortest by the least-squares choice of K). For the moment let us write $\gamma \equiv \gamma(r)$ and note that for specified p, if $\mu = 0$, (4.8) corresponds to the interval:

$$(4.9) \qquad\qquad \bar{\beta}_i \pm \gamma(r) \sqrt{2\sigma^2 \sum_h k_{ih}^2},$$

but, of course, this is *not* the p-level confidence interval for β_i alone, which is in fact:

$$(4.10) \qquad\qquad \bar{\beta}_i + \gamma(1) \sqrt{2\sigma^2 \sum_h k_{ih}^2}.$$

However, it appears that Bienaymé believed that even in considering β_i *alone*, (4.9) should be used in place of the shorter (4.10). This presumably stems from the idea that the variability of *all r estimates* should even then be taken into account. This error in statistical reasoning is presumably of the kind for which Chuprov (1960; cited in our §3.2) takes Bienaymé to task.

This paper of Bienaymé (1852a) is his first on linear least squares. It is additionally important in Bienaymé's career in that it was submitted to the Academy for approval (an abstract appears on p. 458 of *Comptes Rendus* **33,** 1851), and given a very favorable report by Lamé, Chasles, and Liouville (1852), with the recommendation that it be printed in the *Recueil des Savants Etrangers* as eventually happened in 1868, Liouville himself having published it in 1852. The referees say that in regard to the difficult problem he has undertaken, Bienaymé acquits himself *"avec beaucoup d'adresse et de talent,"* and indeed:

> Even quite apart from any idea of appliction to the great questions of natural philosophy, his memoir will be read with interest by mathematicians.

It would appear that, while Bienaymé was not a member of the Academy at the time of the paper's submission, it helped in no small measure in facilitating his election.

In spite of the no doubt adverse effect of Bertrand's criticism, authors such as Czuber (1899) were quite well aware of the value of Bienaymé's startling idea of a simultaneous confidence region and did not fall into the trap of using (4.9) instead of (4.10). Indeed in his §69, Czuber endows Bienaymé's work with important consequences of which there is at best only a suggestion actually present therein, such as the notion of "error ellipses." It is interesting to note that some earlier writers were not as perceptive concerning the difference between (4.9) and (4.10)—see, for example Wrede (1873), and Merriman's (1877) summary of Bienaymé's paper.

4.4. Cauchy's role in interpolation theory

The philosophy of the *interpolational* view of (4.1) is summarized briefly as follows. In physical science, the situation described by (4.1) occurs when it is suspected that for settings $x_1, \ldots, x_r$ there is a value y related by a linear relation:

$$y = \sum_{i=1}^{r} \beta_i x_i,$$

where $\beta = \{\beta_i\}$ is unknown. A number N of observations on y can be made, corresponding to N different sets of $(x_1, \ldots, x_r)$, i.e., we obtain a data set $(Y_i, x_{i1}, x_{i2}, \ldots, x_{ir})$, $i = 1, \ldots, N$, but the readings Y, on y, are subject to error. It is required to approximate β from this information with a view to *predicting* the y value corresponding to any possible set of x values. The least-squares approximation, or estimate, of β, $\hat{\beta}$, is in practice obtained as the solution of the *normal equations:*

(4.11) $$X'X\hat{\beta} = X'Y,$$

which is unique in virtue of the assumption of linear independence of the columns of X, which renders $X'X$ nonsingular. The equations (4.11) arise out of the minimization of $\epsilon'\epsilon = (Y - X\beta)'(Y - X\beta)$ with respect to β. Gauss (1809, 1810) gave an algorithm for solving (4.11) that has become the textbook method (Gaussian elimination) of solving a nonsingular set of linear equations, namely the method of successive elimination of unknowns. This consists of first pivoting successively on diagonal elements to give an upper-triangular array of equations, which are then solved in reverse order, for one unknown at a time, by back substitution. In the case of (4.11), the final effect of the successive eliminations on the matrix $X'X$ is one of multiplying it by a matrix G which is lower-triangular with units on its diagonal, the result being an upper-triangular matrix U with nonzero diagonal elements which are in fact the successive "pivots":

(4.12) $$GX'X = U.$$

   Cauchy (1835a) addresses himself to a new aspect of the interpolation
problem. He points out that if in fact one is concerned with a convergent
series of unknown functions: $\psi_j(x)$, $j = 1, 2, \ldots$, ordered in some
"natural" manner:

$$(4.13) \qquad\qquad y = \sum_{j=1}^{\infty} \beta_j \psi_j(x)$$

with unknown coefficients β_j, $j = 1, 2, \ldots$, it is then of interest (from the
standpoint of both interpolation and of the need to express approximately
the underlying physical law) to know how many terms to retain in the
expansion, to accord with the criterion: that the predicted value Y—on the
basis of the partial expansion with estimated coefficients—should differ
from the "true" value y only by an allowable extent, comparable with
errors committed in making observations.[3] The maximum number of terms
to be conserved, say r ($\geqslant$), and the coefficients β_j, $j = 1, \ldots, r$, are to be
determined on the basis of a (large) number of pairs (of observations with
corresponding x values): (Y_i, x_i), $i = 1, \ldots, N$. The problem manifests the
linear structure (4.1) if we put (as we shall) $x_{ij} = \psi_j(x_i)$. However, one is no
longer, as hitherto, dealing with fixed r but is concerned with considering
successively larger values, until an "adequate" fit obtains. As Cauchy
points out, the (then) known procedures for approaching the problem, such
as the interpolation method of Lagrange—and the same can generally be
said of the Method of Least Squares, depending as it does, on solution of
(4.11)—have the defect that for each r, $r = 1, 2, \ldots$, the estimation
calculations are almost independent of each other, and, rather than becoming easier by virtue of the calculations already done for smaller r, become
increasingly difficult with increasing r.
   Cauchy's own solution to the problem is ingenious in two respects in
particular:

   i. in that, in going from r to $r + 1$, the estimates hitherto obtained for β_1,
      $\ldots, \beta_r$ are not affected;
   ii. in that residuals are automatically calculated at each stage before
      proceeding to the next, enabling one to assess whether the stage reached
      is an adequate stopping point.

We shall now sketch his algorithm; the method is in no way related to the
least-square approach, which does not appear to be mentioned in Cauchy's
memoir at all.
   Suppose for the moment that r terms are to be retained in the expansion.
We operate on the $N \times r$ matrix $X = \{x_{ij}\}$ and the $N \times 1$ vector $\mathbf{Y} = \{Y_i\}$,

---

[3]The situation brings to mind polynomial regression; this aspect is taken up at the conclusion of
§4.5.

to produce a new matrix ΔX and a new vector $\Delta \mathbf{Y}$, as follows. Let $s_i = \mathrm{sign}$ $\{x_{i1}\}$; then $\Delta X = \{\Delta x_{ij}\}$ is $N \times (r - 1)$, $\Delta \mathbf{Y}$ is $N \times 1$, and both are given by:

$$(4.14) \qquad \Delta x_{ij} = x_{ij} - x_{i1} \frac{\sum\limits_{i=1}^{N} s_i x_{ij}}{\sum\limits_{i=1}^{N} s_i x_{i1}}, \qquad j = 2, \ldots, r,$$

$$(4.15) \qquad \Delta Y_i = Y_i - x_{i1} \frac{\sum\limits_{i=1}^{N} s_i Y_i}{\sum\limits_{i=1}^{N} s_i x_{i1}}.$$

In other words, we first form:

$$(4.16) \qquad \begin{cases} \sum\limits_{i=1}^{N} s_i x_{ij} = \sum\limits_{i=1}^{N} x'_{ij} \quad \text{where } x'_{ij} = s_i x_{ij},\, j = 1, 2, \ldots, r, \\ \sum\limits_{i=1}^{N} s_i Y_i = \sum\limits_{i=1}^{N} Y'_i \quad \text{where } Y'_i = s_i Y_i, \end{cases}$$

and use these to eliminate the variable β_1 from the equation system:

$$X\beta = \mathbf{Y}.$$

Note the built-in checks on calculation:

$$(4.17) \qquad \sum_{i=1}^{N} s_i \Delta x_{ij} = 0, \qquad \sum_{i=1}^{N} s_i \Delta Y_i = 0.$$

The estimate for β_1 is then taken as

$$\tilde{\beta}_1 = \sum_{i=1}^{N} s_i Y_i \Big/ \sum_{i=1}^{N} s_i x_{i1} \qquad \left(= \sum_{i=1}^{N} Y'_i \Big/ \sum_{i=1}^{N} |x_{i1}| \right),$$

so that, by equation (4.15), ΔY_i is in fact the residual $Y_i - x_{i1}\tilde{\beta}_1$ from taking as the approximation to (4.13) the first-order expansion $y = \tilde{\beta}_1 \psi_1(x)$.

The new system ΔX, $\Delta \mathbf{Y}$ is then treated in precisely analogous manner to produce an $N \times (r - 2)$ matrix $\Delta^2 X$, and an $N \times 1$ vector $\Delta^2 \mathbf{Y}$, with $\tilde{\beta}_2$ consequently given by a formula analogous to that for $\tilde{\beta}_1$:

$$\tilde{\beta}_2 = \sum_{i=1}^{N} (\Delta Y_i)' \Big/ \sum_{i=1}^{N} |\Delta x_{i2}|.$$

We may regard this step as producing a first-order linear approximation to Δy. The first step of this second stage will have been to form:

$$(4.18) \qquad \sum_{i=1}^{N} (\Delta x_{ij})', \qquad \sum_{i=1}^{N} (\Delta Y_i)'.$$

Other stages follow similarly until the rth stage, when:

$$(4.19) \qquad \tilde{\beta}_r = \sum_{i=1}^{N} (\Delta^{r-1} Y_i)' \Big/ \sum_{i=1}^{N} |\Delta^{r-1} x_{ir}|,$$

the preceding step having been to form:

$$(4.20) \qquad \sum_{i=1}^{N} (\Delta^{r-1} x_{ij})', \qquad \sum_{i=1}^{N} (\Delta^{r-1} Y_i)'.$$

In actual fact r is not given but is to be found. This is achieved as follows: $\tilde{\beta}_1$ and the ΔY_i, $i = 1, \ldots, N$, are calculated as in the first stage above—*these do not depend on $x_{ij}, j \geq 2$*; if all the residuals ΔY_i are small, the process stops with $r = 1$. If not, $\tilde{\beta}_2$ is calculated with $\Delta^2 Y_i$, $i = 1, \ldots,$ N—these calculations do not depend on $\Delta x_{ij}, j \geq 3$, and hence do not depend on $x_{ij}, j \geq 3$; if all the residuals $\Delta^2 Y_i$ are small, we stop with $r = 2$, and $\tilde{\beta}_1$, $\tilde{\beta}_2$ as calculated. The process continues until the stage at which $\Delta^r Y_i$, $i = 1, \ldots, N$, are all small. At most the procedure can be carried on till $r = N$.

It is important to note that in a later paper Cauchy (1847a, Section I) again describes the method but no longer makes absolutely clear what the estimates of β_i are to be; nor does he mention the desirability of the nonchangeability of estimates already obtained. We shall refer to the above method as *Cauchy's Algorithm (or Method) 1.*

Suppose now that, because of the nature of the problem, r *is known a priori.* Then the "first-step" key quantities at each stage, such as (4.16), (4.18), and (4.20), can be written together as:

$$(4.21) \qquad \sum_{i=1}^{N} (\Delta^{t-1} x_{ij})', \qquad \sum_{i=1}^{N} (\Delta^{t-1} Y_i)', \qquad \begin{array}{l} t = 1, \ldots, r \\ j = t, \ldots, r. \end{array}$$

One might then be interested in *an associated* upper-triangular linear equation system:

$$(4.22) \qquad \sum_{j=t}^{r} \left\{ \sum_{i=1}^{N} (\Delta^{t-1} x_{ij})' \right\} \beta_j^* = \left\{ \sum_{i=1}^{N} (\Delta^{t-1} Y_i)' \right\}, \qquad t = 1, \ldots, r.$$

It is to be noted that in general $\beta_j^* \neq \tilde{\beta}_j$, and that the set $\beta_j^*, j = 1, \ldots, r$, will change if r is increased to $r + 1$.

When Cauchy returns to the problem (Cauchy, 1847b, Section I; 1853a) it has become evident to him that his interpolational work of 1835 can be used for the evaluation of unknowns "*determinées par un grand nombre d'équations approximatives du premier degré,*" that is to say, in the case of fixed r and large N. However, *he now takes as the estimator of $\boldsymbol{\beta}$, the vector $\boldsymbol{\beta}^* = \{\beta^*\}$ as determined by the system* (4.22). There is, additionally, one slight difference in setting up the "equations," i.e., essentially

(4.21): At each stage the remaining columns are possibly interchanged to give maximum value among all columns of the sum of the moduli of their elements to the first column. However, quite apart from this, it is clear in particular from p. 1117 of Cauchy (1853a) that an estimation method different from both his Algorithm 1 and least squares is being used. Accordingly, we shall call this *Cauchy's Algorithm 2*. Indeed, if we write (4.22) in matrix notation, we have:

$$(4.23) \qquad CX\beta^* = CY$$

where C is an $r \times N$ matrix of rank r, and CX is upper-triangular. The set of equations giving Algorithm 1's $\tilde{\beta}_j$, $j = 1, \ldots, r$, may be written in related matrix form:

$$(4.24) \qquad BCX\tilde{\beta} = CY,$$

where $B = D(CX)^{-1}$ and $D = \text{diag} \{CX\}$, so that (4.24) is merely $D\tilde{\beta} = CY$. Finally, in view of (4.12), with another choice of C, viz., $C = GX'$, CX reverts to the Gaussian Algorithm's U, and the solution is then the least-squares solution.

While Algorithm 1 and Algorithm 2 have certain similarities, not least of which is the fact that, from (4.17),

$$\sum_{i=1}^{N} s_i \Delta Y_i = 0,$$

and analogously for higher order (that is to say, no matter what the order, r, of regression, the residuals may be adjusted in sign to give zero sum[4]), Algorithm 1 determines the coefficients in a *forward direction and immutably*, while Algorithm 2 is still a *back substitution method*, where, for a given r, the last coefficient of β is determined first and the others thence. It seems clear that Algorithm 2 has little to recommend it as a back substitution method compared to least squares, nor in comparison with the ingenuity of Algorithm 1. In fairness, it must be said that Cauchy (1853a), as well as in following notes to be discussed shortly, appears to still be thinking of Algorithm 1 and its advantages, when in confusion he has factually introduced Algorithm 2. It is interesting that Cauchy takes pains to point out that his new method (Algorithm 2) will give results little different from least squares. Such a point was made earlier by Villarceau (1849b)—see Laurent (1873, pp. 171–172); indeed Cauchy's "method" was arranged in a table convenient for numerical calculation by Villarceau (1849b), the summary of whose paper (Villarceau, 1849a) refers to Cauchy (1847a). However, without recourse to the full memoir of Villarceau, we are unable to determine from secondary sources which of Algorithms 1 and 2 he exposited.

---

[4]Conditions of this sort have been further discussed from a historical viewpoint by Sheynin (1972).

The reader interested in this topic and the following notes on the ensuing controversy between Bienaymé and Cauchy may wish to compare or contrast this account with that given by Seal (1967, §4), who omits mention of Cauchy's (1847a, b, 1853a, b) papers and endows with a beautiful continuity the historical development of least-squares theory, with which, however, the present authors cannot altogether concur.

4.5. Consequences

Cauchy's (1853a) paper was read on 27 June. In a paper read on 4 July, Bienaymé (1853a) takes quick exception to the estimation of β from (4.22) and (4.23), relying heavily on the authority of Gauss and Laplace as regards probabilistic optimality of least squares, as well as its property of minimizing the sum of squares of residuals. He makes the point above of the according inferiority of Algorithm 2 as a back substitution–elimination method also. Unfortunately, it is clear from Bienaymé's paper (apart from a concluding postscript) that he is not addressing himself wholly to Cauchy's (1853a) paper, since he begins with a reference to Cauchy (1835a). Also, on p. 303, Bienaymé asserts that all that really distinguishes Cauchy's method as a new and separate entity is the calculation of $(\Delta^{t-1}Y_i)'$ at each stage; and in this connection he refers to Cauchy's (1835b) paper, which is certainly only concerned with Algorithm 1. It is thus evident from the outset that he, too, is confused between the attributes of Algorithms 1 and 2 and does not distinguish between them. Little wonder, then, that even Merriman's (1877) commentary on Bienaymé (1853a) is guarded.

Cauchy (1853b) replies to Bienaymé's criticism on 18 July by referring at the outset to his memoir of (1835a), and it is to the properties of Algorithm 1 that he alludes constantly, in defending his work against comparison with least squares:

> The algorithm of which I made use in 1835 facilitates this comparison, by reducing the various methods proposed by mathematicians for the resolution of linear equations to several general and very simple formulas, contained in the early pages of my memoir.

Toward the end of the paper, he says that while his new method and least squares both have advantages, his is especially useful in the situation where it is required to find out how many terms to retain (in the regression). Rather mysteriously, he adds that results for the least-squares approach are deducible from his method "*avec une très-grande facilité.*" There is a short reply by Bienaymé at the end of Cauchy's note, to which we shall return. This note of Cauchy increases the confusion and is possibly notable only for his formulation of his work in the simple Δ notation. A final interpolationally oriented paper by Cauchy (1853c) on the affair appears on

25 July. In this he changes stance just a little: He says there are two kinds of problems in interpolation. The first is the case where the number of terms, r, in the regression is fixed *a priori*, for which least squares is suitable. The second is the case it is required to fix r as well as determine the appropriate vector of unknowns, $\boldsymbol{\beta}$; in respect to this, least squares is essentially an "indirect" method, since much calculation is required at each stage, while his method of 1835 (Algorithm 1), is, on the contrary, "direct." He concedes that while "directness" may be possible with least squares, the chief advantage of his own method (Algorithm 1) is the calculation at each stage of residuals $(\Delta^{t-1} Y_i)'$.

It is worth pausing here to elaborate on the issue of "directness" of least squares in Cauchy's sense, since this is certainly an important statistical point, to which we shall return later in this section in connection with Bienaymé and Chebyshev. If we consider a design matrix X_1, produced from X by adding an extra column x, then:

$$X_1'X_1 = \begin{pmatrix} X'X & X'\mathbf{x} \\ \mathbf{x}'X & \mathbf{x}'\mathbf{x} \end{pmatrix}; \qquad X_1'\mathbf{Y} = \begin{pmatrix} X'\mathbf{Y} \\ \mathbf{x}'\mathbf{Y} \end{pmatrix};$$

so if upper-triangularization of $X'X$ has been achieved, relatively little effort is needed to do the same for $X_1'X_1$. Indeed, let us write the resulting set of equations for the $(r + 1)$th stage coefficient vector $\boldsymbol{\beta}(r + 1)$ as:

(4.25)
$$\begin{bmatrix} A & \mathbf{c} \\ \mathbf{0}' & \gamma \end{bmatrix} \boldsymbol{\beta}(r + 1) = \boldsymbol{\delta}(r + 1),$$

where A is $X'X$ upper-triangularized, and $A\boldsymbol{\beta}(r) = \boldsymbol{\delta}(r)$. With self-evident notation, we rewrite (4.25) as:

$$\begin{bmatrix} A & \mathbf{c} \\ \mathbf{0}' & \gamma \end{bmatrix} \begin{pmatrix} \boldsymbol{\beta}^{(r)}(r + 1) \\ \beta_{r+1}(r + 1) \end{pmatrix} = \begin{pmatrix} \boldsymbol{\delta}^{(r)}(r + 1) \\ \delta_{r+1}(r + 1) \end{pmatrix}, = \begin{pmatrix} \boldsymbol{\delta}(r) \\ \delta_{r+1}(r + 1) \end{pmatrix},$$

so that, clearly,

$$A\boldsymbol{\beta}^{(r)}(r + 1) + \beta_{r+1}(r + 1)\mathbf{c} = \boldsymbol{\delta}^{(r)}(r + 1), = \boldsymbol{\delta}(r);$$
$$\beta_{r+1}(r + 1)\gamma = \delta_{r+1}(r + 1).$$

Hence,

$$\boldsymbol{\beta}^{(r)}(r + 1) = A^{-1}(\boldsymbol{\delta}(r) - \beta_{r+1}(r + 1)\mathbf{c},$$

i.e.,

(4.26)
$$\boldsymbol{\beta}^{(r)}(r + 1) = \boldsymbol{\beta}(r) - \beta_{r+1}(r + 1)A^{-1}\mathbf{c},$$

where

$$\beta_{r+1}(r + 1) = \delta_{r+1}(r + 1)/\gamma.$$

Thus, (4.26) expresses the correction which needs to be made to the coefficients $\boldsymbol{\beta}(r)$ of the r-coefficient regression to yield the first r coefficients $\boldsymbol{\beta}^{(r)}(r + 1)$ of the $(r + 1)$-coefficient regression vector $\boldsymbol{\beta}(r + 1)$. Note that $\alpha = A^{-1}\mathbf{c}$ is the solution of the upper-triangular system $A\alpha = \mathbf{c}$. The development just given is a matrix summary of the adjustment method of Goedseels (1902) in going from the r-coefficient to the $(r + 1)$-coefficient stage. He calls it the procedure of Cauchy *(procède de Cauchy)* applied to the Method of Cauchy and the Method of Least Squares. Note the fact that the "correction" description following (4.25) is more generally applicable than just to least-squares determination of β, as Goedseels points out on his p. 150. However, in spite of its title, it is clear that Goedseels's paper has little to do with Cauchy; indeed, insofar as he is specifically referred to at all, it is in connection with Algorithm 2. Basically, what remains of Cauchy is the motivating problem: how to handle economically extension of regression from r to $r + 1$ coefficients. Of some interest also is the fact that in an earlier work Goedseels (1901) actually presents some simplifications to Algorithm 2; but Harter (1974b) erroneously cites the paper in connection with Algorithm 1.

The dispute between Bienaymé and Cauchy takes a personal turn on 8 August with a brief announcement by Bienaymé (1853b) in relation to Cauchy's (1853b,c) communications of 18 and 25 July. We quote it in full (recall that there was already a short reply to the first of these papers):

> There is in the *Comptes Rendus* of the sessions of last 18th and 25th July, an inversion of materials which has led several persons to ask me why I did not reply to the *second* communication of M. Cauchy, which appeared in the *Compte Rendu* of the 25th. Since there has actually been no such *second* communication, and since I desire only to reaffirm my response inserted in the *Compte Rendu* of the session of the 18th, I would very much like it explained how M. Cauchy's memoir, communicated at this session, came ultimately to be split into two publications. One of these, containing precisely the analytical part which our learned colleague had not developed at the session, has been placed in the *Compte Rendu* of the 18th. The other, containing everything actually read on the 18th, and to which above all my response applies, has been inserted in the *Compte Rendu* of the 25th. It seems useful for me to announce clearly that my response, even though published at an earlier time, had been made immediately after the readings of the totality of the memoir.

It is possible that Cauchy finally realized the dichotomy of the problem and arranged the printing of the affair accordingly. However, in what may be a garbled response to these remarks of Bienaymé, Cauchy comments on 29 August—in the discussion on p. 324 to Bienaymé (1853c)—that in one of the notes previously inserted by Bienaymé, Bienaymé demanded an explanation as to how the memoir of Cauchy of *12 August* had been cut in half. [The memoir of this date is actually Cauchy (1853e); it appears that Cauchy is confusing the "cutting" with the appearance of his two papers (Cauchy, 1853e, f), which do say much the same thing.] Cauchy responds that the answer is very simple—the rules relative to printing did not permit its printing in totality (indeed it was mentioned only by title) for the session; and that he regrets it very much. Finally, there is another attack on 8

August by Bienaymé on the paper of Cauchy (1853e), at its conclusion; but discussion of the papers of Cauchy (1853d,e,f,g,h) and Bienaymé (1853c) properly belongs to the context of a probabilistic approach to (4.1), so we defer it to the next section, where the controversy is continued in that setting.

The "Method of Cauchy" as it has come down to us is in fact Algorithm 2. No less a recent commentor than Linnik (1958, Chapter 14, §5) describes it in these terms. Instrumental toward the formation of this modern "understanding" of the "Method" seems to have been the survey of interpolation formulas by Radau (1891), the most relevant parts being his §§13–17 (pp. 328–338), in which §17 gives an account of the Bienaymé–Cauchy argument. Radau's exposition appears studiously to avoid Cauchy's initial motivation, that the r coefficients already determined should not change if the overdetermined system of linear equations needs to be augmented, to go from order r to $r + 1$. Indeed only his very brief §16 deals with the problem of changing order, still judged on size of current $\Delta^r Y_i$ (which, however, no longer have the genuine meaning of "residuals" in the context of Algorithm 2). Curiously, the account of *die Interpolationsmethode von Cauchy* by Burkhardt (1908, §81, pp. 804–823), although heavily influenced by Radau, in fact describes Algorithm 1, giving the estimates $\hat{\beta}_j, j = 1, \ldots, r$, and emphasizing their immutability with increasing r (p. 809):

> This is not so for Cauchy's new method, which will determine the coefficients of the early part of the series independently of coefficients following.

However, even Burkhardt does not distinguish a separate Algorithm 2. Linnik's commentary on Algorithm 2 refers to *statistical* properties of it, but now nevertheless seems the relevant point at which to mention them, since we shall not discuss this algorithm beyond this section. Referring to unpublished thesis work by L. S. Bartnieva (a research student), he demonstrates that a β^* produced by a system (4.23), whether C is "Cauchy's matrix" or not, is *unbiased* for β under the assumption of zero expectation of residuals in (4.1), viz., $E\epsilon = 0$, for then:

$$E\beta^* = (CX)^{-1} CEY = (CX)^{-1} CX\beta = \beta;$$

and he then calculates the efficiency under standard assumptions on ϵ. In the French statistical literature, we find, for example, Risser (1932, pp. 172–184) exposing also Algorithm 2 and, on pp. 201–202, claiming inappropriately, with Cauchy, that the advantage of *this* method is apparent when it is necessary to decide how many terms to take, as well as to affect calculations in an easy fashion. Risser also perpetuates the misapprehension of Bienaymé, that Cauchy:

> . . . has proposed ultimately to correct the result by the method of least squares. . . .

However, Risser's very next statement is very much apropos our further discussion:

> If it is wished to make the procedure conform with the principle of least squares, one is led to the method of Chebyshev. . . .

Indeed, the modern approach to the motivating problem of Cauchy (1835a) is to orthogonalize the (linearly) independent columns of the design matrix X by Gram–Schmidt orthogonalization. This can be done one vector at a time, starting with the first column, and regression coefficient estimates obtained by least squares earlier in the procedure are not affected by consideration of further columns. However, unlike Cauchy's Algorithm 1, this procedure changes the original (natural) form of the regression equations. In fact one successively estimates the entries of a vector γ in a linear system:

$$\mathbf{Y} = Z\gamma + \epsilon,$$

where Z is a matrix with orthogonal columns related to the original design matrix X by $Z = X\bar{U}^{-1}$ where $\bar{U}$ is an upper-triangular matrix. The least-squares estimator of γ, $\hat{\gamma}$, is then related to $\hat{\beta}$ by $\hat{\gamma} = \bar{U}\hat{\beta}$. The first author to make use of an orthogonalized design matrix Z, produced from a given matrix X, was Chebyshev (1855), in a rather special setting in that in $X = \{x_{ij}\}$, $x_{ij} = x_i^{j-1}$, $i = 1, \ldots, N, j = 1, \ldots, q$. In other words, Chebyshev was actually concerned with fitting a polynomial:

$$y = \sum_{j=1}^{q} \beta_j x^{j-1}$$

on the basis of N pairs of observations (Y_i, x_i), $i = 1, \ldots, N$. In this setting, the problem of orthogonalization of X is that of producing from the q powers $1, x, \ldots, x^{q-1}$, a set of q polynomials $T_0(x) = 1, T_1(x), \ldots, T_{q-1}(x)$ which are *orthogonal* in respect to the points $x_1, \ldots, x_N$ in that:

$$\sum_{i=1}^{N} T_s(x_i)T_t(x_i) = 0, \qquad s \neq t,$$

(assuming equal weights). The orthogonal polynomials produced by Chebyshev have come to bear his name, at least for the case where the points $x_i, i = 1, \ldots, N$, are at equidistant intervals; these are widely tabulated. Naturally, he does not use Gram–Schmidt orthogonalization, a later development. He makes clear at the outset of his paper that the coefficients will be produced by a least-squares fit but also points out that he defers the interpolational aspects to later work, intending presently to concentrate on the connection of the problem with continued-fractions theory.

Since the method of orthogonalization achieves Cauchy's original aims within the framework of least squares, it is fitting from a historical viewpoint, and timely in view of the controversy, that Bienaymé chose this article of Chebyshev to translate into French.[5] Indeed, Seal (1967) is of the opinion that the long prefatory footnote by Bienaymé, which mentions the dispute with Cauchy, reveals that Bienaymé understood the full implications of Chebyshev's procedure. While this is not completely evident to us, Bienaymé's manifestly important role as a catalyst in the evolution of interpolational and probabilistic least-squares theory is demonstrated in a sequel by Chebyshev (1859). In the introduction to this, Chebyshev now makes clear with some force that he recognizes the implications, including the ease of calculation in modifying the expression for the residual sum of squares with increasing number of columns in the design matrix, in the process of deciding where to stop the "expansion." Indeed, he states his intention of demonstrating the computational simplicity of his method, in comparison with ordinary nonorthogonalized least squares and Cauchy's interpolation method,[6] even with a fixed number of terms in the regression. He does so in his §7. Cauchy does not emerge in a good light, and it is hard to escape the already mentioned notion that Bienaymé had a not insubstantial role in the motivation of this second paper.

4.6. Bienaymé and Cauchy on probabilistic least squares

The Bienaymé–Cauchy controversy enters a probabilistic phase with the paper of Cauchy (1853d), when he notes at the bottom of p. 160 in relation to a system of overdetermined equations that *while the Method of Least Squares will furnish the most probable results if* ϵ is $\mathfrak{N}(0,\sigma^2 I)$, on the other hand:

> When these conditions are not satisfied, the method of least squares will not provide these values . . . but values which differ appreciably from the most probable ones. This is effectively the conclusion which may be drawn from formulas established in this memoir, as I shall explain in more detail in a following article.

Thus, Cauchy foreshadows his further work on "most probable estimates" in situations where these will differ from the least-squares estimate. One cannot find fault with this aim as a purely theoretical exercise. However, Cauchy is motivated by his earlier interpolational work (specifically Algorithm 2) and initially intends to use his investigations purely as a source of attack, now from the probabilistic side (Bienaymé's defensive standpoint), to show that other algorithms may be optimal. However, in succeeding

---

[5]This translation has been alluded to several times in Chapter 1.

[6]There is no mention of a specific memoir of Cauchy.

communications, he becomes progressively more and more involved with the mathematics for its own sake and forgets his dispute with Bienaymé (at least insofar as can be seen from these papers). Some of his results are important and well ahead of their time from an independent standpoint (quite apart from any relation to Bienaymé), and we shall therefore mention them in this section and the next if only because no satisfactory account of Cauchy's Error Theory exists [the account in Freudenthal (1971) is perhaps the best]. From a practical viewpoint Cauchy's considerations detract little from the primacy of least squares, as we shall see and as Bienaymé was to note.

Before we proceed it is now timely to refer to the model (4.1) where the ϵ_i are i.i.d. with zero mean and finite variance σ^2, as in Gauss's second justification (see §4.2). Referring to the discussion following (4.4), we saw that the least-squares choice (4.3) of K satisfies (4.2) and simultaneously for each $i = 1, \ldots, r$.

$$(4.27) \qquad\qquad \sum_{h=1}^{N} k_{ih}^2 = \min.$$

It may easily be shown that (4.3) is the only matrix to do so, and we shall need this in the sequel.

The paper of Cauchy (1853e) is his first in the area. It is rather involved and heuristic, so before we proceed to an account of it, it may be useful to state at the outset what appears to be the outcome of his deliberations. He appears inherently to show that if the residuals ϵ_i, $i = 1, \ldots, N$, are i.i.d. with symmetric Stable Law with characteristic function (c.f.): exp $\{-c|\theta|^\alpha\}$, $c > 0$, $0 < \alpha \le 2$, then for the most probable estimate in a certain sense defined by him, K must be chosen to satisfy (4.2) and simultaneously for each $i = 1, \ldots, r$,

$$(4.28) \qquad\qquad \sum_{h=1}^{N} |k_{ih}|^\alpha = \min.$$

Thus the least-squares situation where ϵ is $\mathfrak{N}(0, \sigma^2 I)$ emerges in the subcase $\alpha = 2$, from (4.28) and is in accordance with Gauss's first justification. Brief accounts of the paper are to be found in Burkhardt (1908, pp. 814–815) and Freudenthal (1971), which the reader may wish to compare with ours, which we give in modern terms, without introducing modern tools.

In this paper, like Gauss (and in contrast to Laplace and Bienaymé), Cauchy is concerned with a fixed finite N; he assumes at the outset that the residuals are i.i.d. with symmetric density ϕ [$\phi(\epsilon) = \phi(-\epsilon)$], so that its c.f.:

$$\psi(\theta) = \int_{-\infty}^{\infty} e^{-i\theta\epsilon} \phi(\epsilon) \, d\epsilon = \int_{-\infty}^{\infty} \cos\theta\epsilon\phi(\epsilon) \, d\epsilon = 2\int_{0}^{\infty} \phi(\epsilon)\cos\theta\epsilon \, d\epsilon,$$

is real and symmetric. Thus for a linear function $\lambda'\epsilon = \sum_{i=1}^{N} \lambda_i \epsilon_i$ the c.f. is given by:

$$\cdot\Psi(\theta) = \prod_{i=1}^{N} \psi(\lambda_i \theta),$$

which is likewise real and symmetric. Let us suppose $F(x)$ is the distribution function corresponding to Ψ. If both a and b ($b > a$) are continuity points of a distribution function F, then:

$$F(b) - F(a) = \frac{1}{\pi} \int_0^\infty \frac{\mathscr{I}\{(e^{-i\theta a} - e^{-i\theta b})\Psi(\theta)\}}{\theta} \, d\theta,$$

in the improper Riemann sense, where $\mathscr{I}$ signifies the imaginary part (the reader may check from Loève, 1963, p. 188). Since here $\Psi(\theta)$ is real and symmetric, this becomes:

$$F(b) - F(a) = \frac{1}{\pi} \int_0^\infty \frac{(\sin \theta b - \sin \theta a)}{\theta} \Psi(\theta) \, d\theta,$$

which Cauchy is able to write down from Cauchy (1853d); and the formula holds for any b and a since F has a density. Thus, for positive v,

$$(4.29) \qquad P = F(v) - F(-v) = \frac{2}{\pi} \int_0^\infty \frac{\sin \theta v}{\theta} \Psi(\theta) \, d\theta$$

with the integral taken in improper Riemann sense. Cauchy notes also the type of inversion:

$$\phi(\epsilon) = \frac{1}{\pi} \int_0^\infty \psi(\theta)\cos \theta\epsilon \, d\theta,$$

(which we know to hold if, e.g., $\psi(\theta)$ is absolutely integrable) and that:

$$(4.30) \qquad D_v P = \frac{dP}{dv} = \frac{2}{\pi} \int_0^\infty \Psi(\theta) \cos \theta v \, d\theta$$

(which we know to be valid under a similar condition). For the rest, Cauchy's analysis is quite heuristic, and we do not attempt to justify it.

Cauchy now considers an arbitrary K satisfying (4.2) and takes λ' to be its first row. Note that in view of the fact that the ϵ_i are symmetrically distributed and i.i.d., the distribution of $\Sigma \lambda_i \epsilon_i$ is also that of $\Sigma |\lambda_i| \epsilon_i$, so we will take without loss of generality $\lambda > 0$ in the consideration of $\Sigma \lambda_i \epsilon_i$. *He is now concerned with maximizing the probability P, first for fixed v, under variation of λ* [under the constraint (4.2)]; *then with imposing the further condition that the optimization be independent of v.* Taking for the moment v as fixed, then *at the maximum* the perturbation in P, δP, corresponding to a perturbation in λ, satisfies $\delta P = 0$; and since the point is to be further invariant in respect of changes of v,

(4.31) $D_v \delta P = 0.$

From (4.30), $D_v P/2$ is the density corresponding to $\Psi(\theta)$, so it follows that:

$$\Psi(\tau) = \int_0^\infty \cos v\tau \cdot D_v P \, dv,$$

whence from (4.31) it is plausible that:

(4.32a) $\delta\Psi(\tau) = \int_0^\infty \cos v\tau \cdot D_v \delta P \, dv = 0,$

for any value of the parameter τ. Consequently,

(4.32b) $\delta\Psi(1) = 0.$

Now put:

(4.33) $w(\theta) = \psi'(\theta)/\psi(\theta) = d \log \psi(\theta)/d\theta;$

then taking logs and perturbing in $\Psi(\theta) = \Pi\psi(\lambda_i\theta)$,

$$\frac{\delta\Psi(\tau)}{\Psi(\tau)} = \tau \sum_{i=1}^N w(\lambda_i\tau)\delta\lambda_i,$$

so from (4.32),

(4.34a) $\displaystyle\sum_{i=1}^N w(\lambda_i\tau)\delta\lambda_i = 0,$

(4.34b) $\displaystyle\sum_{i=1}^N w(\lambda_i)\delta\lambda_i = 0.$

At this point Cauchy's argument passes from heuristic to obscure; certainly (4.34a) will be consistent with (4.34b) if $w(\lambda\tau)/w(\lambda)$ is independent of λ (which he asserts must necessarily be so). Then:

$$w(\lambda\tau)/w(\lambda) = w(\tau)/w(1), \qquad \tau, \lambda > 0.$$

It follows (and we would now use a well-known argument involving the Cauchy functional equation) that, if $w(\lambda) > 0$ for $\lambda > 0$,

$$w(\lambda\tau)/w(\lambda) = \tau^\rho \qquad \text{for } \tau > 0,$$

for some ρ, $-\infty < \rho < \infty$, whence:

$$w(\tau) = \tau^\rho w(1), \qquad \tau > 0.$$

Thus, recalling the definition of w in (4.33),

$$\log \psi(\tau) = \frac{w(1)\tau^{\rho+1}}{\rho + 1} + \text{const.}, \qquad \tau > 0,$$

i.e.,

$$\psi(\theta) = \exp\{-c\theta^\alpha\}, \qquad \tau > 0,$$

where $\alpha = \rho + 1$, $c = -w(1)/(\rho + 1)$, and the arbitrary constant has been removed by the condition $\psi(0) = 1$, which also implies $\alpha \geq 0$; and the fact that $\psi(\theta) \leq 1$ (by putting $\theta = 1$) implies $c \geq 0$. To exclude trivialities one needs take $\alpha > 0$, $c > 0$. Then, by symmetry of ψ:

(4.35) $$\psi(\theta) = \exp\{-c|\theta|^\alpha\}, \qquad \text{for all } \theta,$$

and the corresponding density by the inversion mentioned earlier is:

(4.36) $$\phi(\epsilon) = \frac{1}{\pi} \int_0^\infty \exp\{-c\theta^\alpha\} \cos\theta\epsilon \, d\theta.$$

Let us now return to the original problem, assuming the common density of residuals is given by (4.36); we require an $r \times N$ matrix K satisfying $KX = I$, which, if it is to yield the most probable estimate in the sense of Cauchy, will *simultaneously maximize* [from (4.29)] for each $i = 1, \ldots, r$,

$$P_i = \frac{2}{\pi} \int_0^\infty \exp\{-s_i c\theta^\alpha\} \frac{\sin\theta v}{\theta} \, d\theta,$$

where:

$$s_i = \sum_{h=1}^N |k_{ih}|^\alpha,$$

since:

$$\Psi(\theta) = \exp\{-s_i c\theta^\alpha\},$$

where we are considering the ith row of K. In other words, for this particular common distribution of residuals, Cauchy's criterion requires a K, which, subject to (4.21), simultaneously minimizes (4.28) for $i = 1, \ldots, r$, as we asserted earlier.

Cauchy notices directly from performing the integration (4.36), that setting $\alpha = 2$ yields an $\mathfrak{N}(0, 2c)$ density, while setting $\alpha = 1$ yields the (so-called Cauchy) density:

$$\phi(\epsilon) = \frac{c}{\pi(c^2 + \epsilon^2)}, \qquad -\infty < \epsilon < \infty.$$

We know *now* that (4.35) with $c > 0$ and $0 < \alpha \leq 2$ is the *general form* of the characteristic function for the nondegenerate symmetric Stable Laws and, since (4.35) is absolutely integrable over $(-\infty,\infty)$, has continuous density (e.g., Lukacs, 1960, pp. 102–103). At the end of this memoir and in Cauchy (1853f), he speaks also of densities corresponding to (4.35) with $\alpha > 2$, which is thus not a characteristic function, for if it were, it too would correspond to a symmetric Stable Law. Moreover Czuber (1899, §61) later makes the same error. Indeed this paper of Cauchy is perhaps the first in which Stable Laws make an appearance and is also historically important because of its connection with various inversion formulas and the Cauchy functional equation.

There is a brief rejoinder by Bienaymé on p. 206 of Cauchy (1853e) that the examination of questions as delicate as those treated by Cauchy would not be fruitful in a verbal discussion. Anticipating Bienaymé (1853c), he believes he can bring forth several arguments in support of the opinion of Laplace (*à l'appui de l'opinion de Laplace*) that the method may validly be applied *irrespective of the probability law governing the residuals*. He is evidently thinking, as becomes clear from the 1853c paper, of the *asymptotic theory* as $N \to \infty$; but insofar as Laplace's considerations require the existence of moments of residuals for their development, a property not enjoyed in general by the symmetric Stable Laws, these objections are not completely relevant, all the more since Cauchy's considerations are for fixed N and not aimed at Laplace. However this brief response may have elicited from Cauchy (1853f) a consideration of what may happen when $N \to \infty$. This follow-up note is divided into two not well-connected parts, the second of which only reiterates the considerations of the preceding paper. [It has been mentioned in §4.5 that the printing of the two notes (1853e and 1853f) may have been inverted.]

The first part concerns itself with rather heuristic considerations indicating the kind of conditions under which it might be expected that the error of estimate:

$$\lambda'\epsilon = \sum_{i=1}^{N} \lambda_i \epsilon_i,$$

would be approximately normal for large N, presumably seeking to explain the degree of applicability of Laplace's asymptotic justification for least squares. Cauchy assumes the common symmetric density, $\phi(\epsilon)$, of residuals is concentrated on a finite interval $[-\tau,\tau]$, and the common variance is denoted by σ^2. (Note that $E(\lambda'\epsilon) = 0$, Var $\lambda'\epsilon = \sigma^2 \sum_{i=1}^{N} \lambda_i^2$). He remarks that if $\psi(\theta)$ and $\Psi(\theta)$ have the prior meanings, $\psi(\theta) \approx \exp -\zeta\theta^2$ for *small* θ, where ζ depends on θ but is close to $\sigma^2/2$, then, if all $\lambda_i\theta$, $i = 1, \ldots, N$, are small,

$$\Psi(\theta) = \prod_{i=1}^{N} \psi(\lambda_i\theta) \approx \exp -\left(\left\{\sum_{i=1}^{N} \lambda_i^2\right\} \frac{\sigma^2}{2}\right),$$

and discusses heuristic circumstances for this to be increasingly accurate.

As we now know, consideration of the asymptotic distribution of $\sum_{i=1}^{N} \lambda_i \epsilon_i$ as $N \to \infty$ amounts to a Central Limit Theorem type of result for the sum of N independent but not identically distributed r.v.'s $\{\lambda_i \epsilon_i\}$, $i = 1, \ldots, N$; and that if we assume the λ_i to be uniformly bounded for all i, we can apply as $N \to \infty$ a well-known theorem (on inhomogeneous sums) the first version of which is due to Chebyshev (1887), as Sleshinsky (1892), in his preamble to a rigorization of Cauchy's Method, points out. Ironically, as we have mentioned elsewhere,[7] Chebyshev based his proof on methods derived from Bienaymé's 1853c paper. This theorem (see, e.g., Fisz, 1963, pp. 206–207) implies that a necessary and sufficient condition for the distribution of:

$$(4.37) \qquad \sum_{i=1}^{N} \lambda_i \epsilon_i \bigg/ \left\{ \sigma^2 \sum_{i=1}^{N} \lambda_i^2 \right\}^{1/2} \to \mathfrak{N}(0,1),$$

as $N \to \infty$, is:

$$(4.38) \qquad \sum_{i=1}^{\infty} \lambda_i^2 = \infty.$$

If the λ_i's themselves are permitted to depend on N so that each λ_i itself is (of order) N^{-1}, as Cauchy suggests, (4.37) again obtains under a condition similar to (4.38) by the same theorem, after elimination of N in the ratio of (4.37) between numerator and denominator.

Bienaymé's ultimate views on the subject occur within the context of the famous 1853c paper. This paper is intended as an eloquent and impassioned defence of Laplace's asymptotic approach to linear least-squares theory with arbitrary symmetric residual distribution, for whose full flowering Bienaymé (1852a) was himself responsible and the exactitude of which Cauchy has sought to deny in his writings. However, he says (in a typically verbose and polemical introduction aimed at Cauchy), far from denying the validity of the Laplacian approach in Cauchy's writings:

> . . . there are parts of this so fruitful and sometimes very ingenious analysis which, with little change, clearly demonstrate the discovery of Laplace.

We have seen this to be so in Cauchy (1853f); and perhaps this indeed was the purpose of its first part; but Bienaymé evidently takes it as a criticism.

Before he reverts to his defence in this specific matter, Bienaymé makes some interesting remarks which are worth noting in that they reveal his personal philosophy concerning science in general, to him accordant with the sacred duty in defending Laplace. He regards (his own) criticism not as an end in itself, but as a tool of which he is obliged to make use to determine the truth, specifically in an area such as the calculus of probabilities which, even among the illustrious, is susceptible to giving rise to illusions. Impar-

---

[7]See §5.10.

tial criticism is essential to the actual progress of science; in seeking the truth of things one is forced to clear and then *defend* the terrain, so successors may build thereon with confidence. He intends therefore that his future communications will be of a nature commensurate with these ideas. His own researches on the matter in hand go back a long time, and his demonstrations of certain errors go back 30 years, he claims. This long silence should make it evident that he has no desire to criticize *per se*.

We may wonder in retrospect whether Bienaymé's early reputation in statistical theory as elucidator was not in part of his own creation through such statements, whereas the central truth of the matter is an almost *obsessive* defence of Laplace against attacks real and imagined, valid or not. For example, reverting now to the particular, he states that the opinion of Laplace is worthy of every defence, for if Laplace's opinion on least squares in the current controversy with Cauchy is not strictly correct, the best part of his treatise of 1812 will be invalidated, since a number of related chapters therein will also be cast into doubt. Bienaymé claims that the central idea of Laplace, underlying also the least-squares theory, is effectively the notion of a limit theorem (as we would now call it), in the sense that a "universal" specific limit distribution reduces many specific cases to a single one if the sample size is large. Thus, implicitly, by attacking Laplace's approach to least squares, one attacks this unifying principle of Laplace's work, in Bienaymé's thinking. This stance is conservative, incorrect, and emotional; associated with it is his implicit refusal to allot any value to *small sample theory* (N fixed) which underlies both Gauss's considerations and Cauchy's to this point, largely, it seems, on the grounds that Laplace paid it little attention. On the other hand, it is quite true, as Bienaymé says, that from a *practical viewpoint* the rigorous embellishments of Poisson, Cauchy, and the rest detract not at all from the applicability of Laplace's results.

The sequel of Bienaymé's paper contains some perspicacious comments and results in the calculus of probabilities. Although they have come down to us largely dissociated by time from Bienaymé's name, these form the true scientific content of the paper. Their association with the motivation for the paper appears, at least in our time, tenuous but we shall develop them from this viewpoint insofar as possible. Bienaymé begins his *mathematical* exposition with the two components which he considers distinguish Laplace's asymptotic results:

i. the standard [Normal] error integral appearing as limit law; and
ii. the presence in asymptotic expressions of the mean-square difference between each possible error and its mean; or, as we would now say, the *variance of the error*.

After making some eminently sensible statements *re* i, dispelling the misconception (arising from its ubiquitous appearance in probability calculus) that the Normal is *the* Error Law in some sense, he goes on to ii to explain why the *variance* plays a fundamental and probabilistic—rather than analytical—part in probability calculus and why, indeed, it cannot be replaced by another arbitrary measure of precision, such as any even moment about the mean.

To expand on the last point: Bienaymé uses *generating functions* to prove, if $X_1, X_2, \ldots, X_N$ are independent random variables (he assumes them discrete, and independence, as usual for the time, is inherent in stating that the generating function for a sum of r.v.'s is the product of the individual generating functions), the conservative property of the mean:

$$E\left(\sum_{i=1}^{N} X_i\right) = \sum_{i=1}^{N} EX_i,$$

which, as he states, is well known, and goes on to show the same for the variance:

$$\text{Var}\left(\sum_{i=1}^{N} X_i\right) = \sum_{i=1}^{N} \text{Var } X_i,$$

which identity now, occasionally, is called the *Bienaymé Equality*. He further points out (if we write $\mu_i = EX_i$), that while it is still true that:

$$E\left(\left\{\sum_{i=1}^{N} X_i - \sum_{i=1}^{N} \mu_i\right\}^3\right) = \sum_{i=1}^{N} E(\{X_i - \mu_i\}^3),$$

for the fourth power the conservative property fails, for:

$$E\left(\left\{\sum_{i=1}^{N} X_i - \sum_{i=1}^{N} \mu_i\right\}^4\right) - 3E^2\left(\left\{\sum_{i=1}^{N} X_i - \sum_{i=1}^{N} \mu_i\right\}^2\right)$$
$$= \sum_{i=1}^{N} \{E(\{X_i - \mu_i\}^4) - 3E^2(\{X_i - \mu_i\}^2)\}.$$

Indeed, if $X_1, X_2, \ldots, X_N$ are i.i.d. (as X, say), then as $N \to \infty$:

$$E^{1/2}\left(\left\{\sum_{i=1}^{N} X_i - \sum_{i=1}^{N} \mu_i\right\}^4\right) \sim \sqrt{3}N \text{ Var } X,$$

so one cannot avoid discussion of variance as a measure of dispersion by using instead the fourth central moment about the mean. He shows the same to be true as $N \to \infty$ for the $2i$th moment:

$$E^{1/i}\left(\left\{\sum_{i=1}^{N} X_i - \sum_{i=1}^{N} \mu_i\right\}^{2i}\right) \sim \text{const. } N \text{ Var } X,$$

while also recognizing that odd central moments are of no utility in discussing dispersion. He concludes this part of the discussion by noting that if ϵ_i are i.i.d.r.v.'s (distributed as ϵ, say) then:

$$\text{Var } \lambda'\epsilon = \lambda'\lambda \text{ Var } \epsilon = \left(\sum_{i=1}^{N} \lambda_i^2\right) \text{Var } \epsilon,$$

which could clearly lead onto a discussion of *least squares itself* to minimize the variance, but he does not proceed in this direction explicitly.
Stigler (1974) is of the opinion that by these arguments Bienaymé sought to
show that for large samples the Method of Least Squares is best linear for
any even loss function. This is an extrapolation which is neatly consistent
with the aims of Bienaymé's paper. However, it *is* generally true that this
discussion of Bienaymé has had influence on the choice of second central
moment as measure of dispersion even in practical settings; thus in Chebotarev's (1961) brief Russian survey of the history of least squares:

> Even so, the problem of a firmer foundation [for the second central moment] . . .
> remained inadequately resolved until the learned Frenchman, Bienaymé,
> deduced that description of dispersion of random errors was appropriate only by
> . . . second powers. . . .

Bienaymé's second argument in relation to the variance as a measure of
dispersion centers around the Bienaymé–Chebyshev Inequality, which
merits a separate section.[8]

In the last few pages of his paper Bienaymé hastens to come to grips with
the exceptions indicated by Cauchy (that is to say, the symmetric Stable
Laws with index α, $0 < \alpha < 2$, excluding the Normal distribution $\alpha = 2$).
They do not at all modify the opinion of Laplace, he says. Actually, his
remarks appear, understandably, to pertain mostly to the case $\alpha = 1$, of the
now so-called Cauchy distribution—already alluded to in the work of
Poisson (1824)[9]—where he takes $c = 1$ in (4.35) to obtain the density:

$$\phi(\epsilon) = \frac{1}{\pi(1 + \epsilon^2)}, \qquad -\infty < \epsilon < \infty,$$

although he may well have been aware that Var ϵ is infinite for the other
non-Normal cases also. He makes the point (which we have noted earlier)
that in any practical setting with proper instruments, errors will have finite
moments, and such pathological densities as the Cauchy cannot arise
without the experimenter being aware of the situation. His final remark
about the Cauchy distribution is that if one has N i.i.d.r.v.'s with this
distribution, then their sample mean:

$$\bar{\epsilon} = \sum_{i=1}^{N} \epsilon_i / N,$$

still has the same distribution, and so is no more "accurate" than a single
measurement, and it would indeed be a singular instrument which possessed such a property. All this is said somewhat in the spirit of ridiculing

---

[8]See §5.10.
[9]See Stigler (1974) and our §2.5.

Cauchy, against the previously expressed intention of constructive criticism.

There is a response by Cauchy which appears on pp. 324–326 of the *Comptes Rendus* version of Bienaymé (1853c) but is not included in the 1867 version of it. Here Cauchy states that his work and Bienaymé's are from a practical view of great interest to astronomers and geometers (mathematicians). He further says (distorting what Bienaymé has actually said) that, as Bienaymé has remarked, it is possible to derive from his (Cauchy's) calculations a more exact and rigorous justification of least squares. What he (Cauchy) has shown, furthermore, are the relative advantages and disadvantages of various methods, particularly least squares, and when they are applicable.

This stage marks the end of the controversy, at least in printed manifestation. Continuing bitterness, however, is nevertheless suggested by Bienaymé's (1871) note[10] in defence of Cournot. It is possible that Bienaymé felt that Cauchy had really had, in view of his great prestige, the majority of supporters in the interpolational part of the controversy, even though, as we have seen, it was Cauchy himself who was responsible for that confusion. On the probabilistic side of the controversy, we have tried to demonstrate at various points that the two men argued at cross purposes, both being correct. [In this we agree with Stigler (1974).] The paper of Bienaymé (1853c) does not, therefore, constitute a refutation of Cauchy's work. Indeed, for this reason, it probably created, in its primary aim, more confusion than light: in, e.g., Laurent (1873, p. XI) we have:

> I believe myself to have exposited the method of least squares according to the manner indicated by Cauchy, but with the modifications which M. Bienaymé has published in his celebrated memoir . . . of which, because of its great scope, I have not found it possible to give an analysis.

and in Merriman (1877) and Czuber (1899) only very brief allusions to Bienaymé's paper. Likewise, the remarks of the next section notwithstanding, Cauchy's writings on the probabilistic side of the controversy have been almost completely and unjustifiably eclipsed; one opinion, unfortunately justified by their obscurity, is Freudenthal (1971):

> This was a muddy chapter of Cauchy's work. . . .

4.7. Cauchy continues

We have already mentioned several of Cauchy's remarks in apparent reply to Bienaymé (1853c). However a reply, *per se*, is not his aim. He says that he has replaced the hitherto inexact analysis relating to the asymptotic

---

[10]See §1.4.

discussion of least squares by rigorous and exact formulas. This work was principally contained in a memoir read at the last session of the Academy and which he intended to present at the current session (presumably these eventually became Cauchy, 1853g, 1853h). He requests that in view of the subject's importance the Academy order the printing of the first of these under the current session (22 August). As a relevant aside we find that Chevreul spoke against this proposition, along the lines that if the *Comptes Rendus* were to satisfy a real need for the rapid dissemination of scientific information, it could not be a forum for long memoirs and discussions, but only for the essence of the first and résumés of the second. In accordance with such aims, the rules governing publication stated that a member could not have at his disposal within 1 year more than 50 pages in all, nor more than 8 in a single number. Evidently Cauchy's current request was not in line with this—indeed the limit on publication had been instituted to restrain him in the first place (Grattan-Guinness, 1970, p. 27). However, it manifests Cauchy's influence and persistence that the Academy agreed to his request, even though a placatory proviso stated the case was not to be regarded as a precedent.

In point of fact the paper of Cauchy (1853g) has nothing to do with asymptotics as $N \to \infty$ or with least squares but is a return to his preoccupation with demonstrating situations in which least squares is not optimal. This very brief note appears to be totally unknown but presents a statistically interesting estimation criterion which does not appear to have been explored since and certainly provides a more interesting situation of when least squares is not optimal than hitherto. As usual we are interested in choice of K to satisfy (4.2), supposing the ϵ_i are i.i.d., with symmetric distribution whose range is $[-\tau, \tau]$, $0 < \tau < \infty$. The error of estimate of β by $\hat{\beta}$ is $K\epsilon$; so if we denote the ith row of K by $\mathbf{k}_i'$, the error of estimate in $\hat{\beta}_i$ is $\mathbf{k}_i'\epsilon$, where $\mathbf{k}_i'$ is to be chosen in some optimal manner under the constraint $\mathbf{k}_i'X = \mathbf{f}_i'$, where $\mathbf{f}_i$ is the vector with unity in the ith position and zeroes elsewhere. The maximum attainable error of estimate of β_i is therefore:

$$\tau \sum_{h=1}^{N} |k_{ih}|,$$

and Cauchy's optimality criterion is to chose $\mathbf{k}_i$ under its constraint to minimize this. That is, K must be chosen to satisfy (4.2) and (4.28) with $\alpha = 1$, i.e., for each $i = 1, \ldots, r$,

$$\sum_{h=1}^{N} |k_{ih}| = \min.$$

He resolves the problem of finding K in the case $r = 1$ and discusses rather loosely the case $r = 2$.

We may formulate the problem within a general mathematical setting as follows. For a fixed known vector $\mathbf{c}$ and an $r \times N$ matrix ($N \geq r$) of full row rank r, A, it is necessary to *find a solution* $\mathbf{y} = \{y_i\}$ *to the linear equation set:*

$$A\mathbf{y} = \mathbf{c},$$

which minimizes:

(4.39)
$$\sum_{i=1}^{N} |y_i|.$$

Cauchy's procedure for resolving the problem can be explained as follows. There is at least one $r \times r$ submatrix A_1, formed from r suitably selected columns of A, which is nonsingular. Suppose for ease of exposition that the first r columns of A can be so selected; then one solution to $A\mathbf{y} = \mathbf{c}$ is well known to be:

$$\mathbf{y}^* = \begin{pmatrix} \mathbf{y}_1 \\ \mathbf{0} \end{pmatrix} \quad \text{where } \mathbf{y}_1 = A_1^{-1}\mathbf{c}.$$

Cauchy appears to suggest that one need search only among solutions of type $\mathbf{y}^*$, obtained by means of nonsingular $(r \times r)$ matrices, to minimize (4.39). A crude upper bound for the number of these solutions is $\binom{N}{r}$, and each solution contains at least $N - r$ zero entries. This proposition can be verified by modern tools quite readily by first reformulating the problem in standard preliminary linear-programming format, so that elementary ideas of that theory (e.g., Gass, 1958, Chapter 3) become applicable. The verification, together with that for a problem where A is not of full row rank, is given by Seneta (1976). The reader should note that the problem differs, in spite of certain common features, from those treated by Wagner (1959), W. D. Fisher (1961), and Davies (1967).

The general purpose, on the other hand, of Cauchy (1853h) is to amplify on the remarks made in Cauchy (1853f), and so is very much to the point apropos Laplace's (first) asymptotic justification of least squares, which has been the subject of Bienaymé's defence. However, Cauchy begins the paper by remarking that in view of the ruling in respect of publication in the *Comptes Rendus*, he still cannot present his work in its entirety and will confine himself to summarizing the principal results. Some time is spent on recalling the transform and inversion formulas of Cauchy (1853e), and also, on the *"propriétés remarquables"* of characteristic functions corresponding to symmetric densities, which he takes *to be confined to a finite interval* $[-\tau, \tau]$. Returning to the exposition of our previous section, Cauchy then concerns himself in essence with estimating P given by (4.29) with the aid of appropriate truncated integrals, such as:

$$\frac{2}{\pi} \int_0^\Theta \frac{\sin \theta v}{\theta} \Psi(\theta) \, d\theta$$

as $N \to \infty$, where the λ_i's and Θ are permitted to depend on N, to clarify his earlier commentary.[11] We reproduce two of the estimates stated here to give the flavor, as (4.40) and (4.41):

$$(4.40) \qquad I_1 = \left| \frac{2}{\pi} \int_{\Theta}^{\infty} \frac{\sin \theta v}{\theta} \Psi(\theta) \, d\theta \right| \leq \frac{1}{\pi \mathfrak{N}} \exp - \mathfrak{N}$$

where:

$$\mathfrak{N} = \frac{1}{2} \frac{r \Lambda \Theta^2}{1 + r \lambda^2 \Theta^2},$$

with $\lambda = \max (\lambda_1, \ldots, \lambda_N)$, $\Lambda = \boldsymbol{\lambda}' \boldsymbol{\lambda} = \Sigma \lambda_i^2$, $r = \rho(0+) = 2c$, with the function $\rho(\theta)$ defined from the characteristic function $\psi(\theta)$ by $\psi^2(\theta) = \{1 + \theta^2 \rho(\theta)\}^{-1}$ (that is, r is actually the *variance*, σ^2):

$$(4.41) \qquad I_2 = \left| \frac{2}{\pi} \int_0^{\Theta} e^{-c \Lambda \theta^2} \frac{\sin \theta v}{\theta} \, d\theta - \frac{2}{\pi} \int_0^{\Theta} \Psi(\theta) \frac{\sin \theta v}{\theta} \, d\theta \right|$$

$$\leq \frac{2h\sqrt{3}}{\pi} \log \left(\frac{\Theta v}{\sqrt{3}} + \sqrt{1 + \frac{\Theta^2 v^2}{3}} \right),$$

h being the larger of:

$$(4.42a) \qquad \exp \{ c \Lambda \lambda^2 \Theta^4 r^2 \} - 1, \qquad 1 - \exp \left\{ - \frac{c^2 \Lambda \lambda^2 \Theta^4}{1 + c \lambda^3 \Theta^2} \right\}.$$

Thus, if we suppose that Θ is of order[12] exceeding $N^{1/2}$ but less than N, and each λ_i is uniformly of order N^{-1}, then $\mathfrak{N} \to \infty$ with N, so (4.40) becomes increasingly small; while if in fact Θ is of order less than $N^{3/4}$, then (4.41) also approaches zero as $N \to \infty$. Thus the Normal approximation is expected to become increasingly accurate. (In such circumstances least squares is optimal according to the justification of Laplace.)

In actual fact, in view of the lack of proofs in Cauchy's paper, we are fortunate to have a follow-up to it by Sleshinsky (1892), who notes that while the Chebyshev theorem can be used to prove asymptotic Normality of (4.37) its own proof of 1887 also "cannot be called simple" [it is, as we now know, incomplete, and is corrected by Markov (1898) somewhat later], and it is therefore of interest to turn to the manner of Cauchy's investigations of the problem. Pertinently, he points out that Cauchy's investigations of probability are neglected. Following a long Introduction, Sleshinsky embarks on a carefully detailed and rigorous reworking of

---

[11]It is useful to bear in mind that without loss of generality we may assume $\lambda > 0$, as remarked earlier.

[12]We say here $a(N)$ is of order $b(N)$ if there exist constants α, β such that $0 < \alpha < |a(N)/b(N)| < \beta$, all N.

Cauchy (1853h), following certain hints and indications in the earlier Cauchy memoirs. Proofs are given of Cauchy's auxiliary assertions, then of (4.40) and (4.41), under appropriate assumptions about the size of N and Θ, and then of (4.41) with (4.42a) modified to:

(4.42b) $\exp\{c\Lambda\lambda^2\Theta^4\tau^2/4\} - 1$, $1 - \exp\left\{-\dfrac{c^2\Lambda\lambda^2\Theta^4}{1 - c\lambda^2\Theta^2}\right\}$,

if $\Theta < 1/\lambda\tau$. There is also a new estimate (p. 249):

(4.43) $I_3 = \left| \dfrac{2}{\pi} \displaystyle\int_0^\Theta e^{-c\Lambda\theta^2} \dfrac{\sin\theta v}{\theta} \, d\theta - \dfrac{2}{\sqrt{\pi}} \int_0^{v/[2(c\Lambda)^{1/2}]} e^{-x^2} \, dx \right| \leq \dfrac{e^{-c\Lambda\Theta^2}}{\pi c\Lambda\Theta^2}$,

and finally the result (p. 250) that (recall ϵ is confined to $[-\tau,\tau]$):

$$\sum_{i=1}^N \lambda_i\epsilon_i \left/ \left\{\sigma^2 \sum_{i=1}^N \lambda_i^2\right\}^{1/2} \right. \to \mathfrak{N}(0,1),$$

as $N \to \infty$, provided that the numbers:

$$|N\lambda_1|, |N\lambda_2|, \ldots, |N\lambda_N|,$$

are uniformly bounded away from 0 and ∞ (which agrees with our earlier remarks). There follows a general discussion of the result and its relation to least squares.

The reader should note that since (4.40), (4.41), and (4.43) yield, by the triangle inequality, that:

$$\left| \dfrac{2}{\pi} \int_0^\infty \dfrac{\sin\theta v}{v} \psi(\theta) \, d\theta - \dfrac{2}{\sqrt{\pi}} \int_0^{v/[2(c\Lambda)^{1/2}]} e^{-x^2} \, dx \right| \leq I_1 + I_2 + I_3,$$

Cauchy's procedure is a manifestation of estimation of closeness of two distribution functions with the aid of characteristic functions, a topic of subsequent importance in probability theory, whose origins have hitherto been traced to Liapounov (1901). Indeed, in a preceding paper, Liapounov (1900) cites Sleshinsky (1892) and mentions Cauchy. It is notable also that the condition that the λ_i be of order N^{-1} uniformly in i does not do justice to the actual estimates and may be substantially relaxed. Freudenthal (1971) remarks, rather mysteriously, and certainly without reference to Sleshinsky, that:

> The present author adheres to the heterodox view that Cauchy's proof [of the Central Limit Theorem] was rigorous even by modern standards.

The degree of validity of this statement is now evident from Sleshinsky's work. For its time, with its rigor and difficult and involved estimates, it represents an important and neglected epoch in the evolution of probability theory. Indeed, one of its motives, as Sleshinsky implies in his Preface, is to bring to bear the rigor of mathematical analysis on probabilistic propositions, which at the time (with the possible exception of some of Chebyshev's writings) was lacking, with evident and unsatisfactory consequences. The criticism of lack of rigor is valid for all of Laplace, Poisson, Cauchy, and Bienaymé; to read Sleshinsky after attempting to extract the essence of those authors is to come face to face, for the first time, with modern probability theory.

Sleshinsky's paper is also notable, as has been mentioned a number of times before, for a careful commentary on the relationship between Bienaymé, Cauchy, and Chebyshev (see especially pp. 201–210), and it is thus fitting to conclude this chapter with his own assessment of the Bienaymé–Cauchy controversy:

> It became evident from the results, that the truth was on Bienaymé's side, and Cauchy hastened to arrive at it, while exposing to doubt previous proofs and applying new approaches.

5. Other probability and statistics

There is no national science just as there is no national multiplication table; what is national is no longer science.

A. Chekhov

5.1. Introduction

The previous three chapters are concerned with the cases in which Bienaymé has written a group of papers around a theme, or in a specific area. Those of his contributions in probability and statistics that do not fit within such a framework are discussed in this chapter. Other miscellanea are discussed in Chapter 6.

From a modern point of view, many of Bienaymé's most important contributions occur in relative isolation (such as his branching process contribution) or buried in other material and incidental to its aims (such as the work relating to sufficiency and the Bienaymé–Chebyshev Inequality). We take each of the papers, or such identifiable concepts, in chronological sequence and give details on both the contribution and its place in the history of the subject.

The sections in this chapter are, in order: a limit theorem, medical statistics, the law of averages, electoral representation, the concept of sufficiency, a general inequality, a historical note on Pascal, the simple branching process, the Bienaymé–Chebyshev Inequality and, finally, a test for randomness. They well exhibit the breadth of Bienaymé's scientific contributions.

5.2. A limit theorem in a Bayesian setting

In this section we concern ourselves with a probability limit theorem proved (but not quite rigorously) by Bienaymé (1838b) and with its significance and his motivation for it.

A bare mathematical formulation of it is as follows:

Let the random variables Θ_i, $i = 1, \ldots, m$, satisfy $0 \leq \Theta_i \leq 1$, $\Sigma \, \Theta_i = 1$, and the joint probability density function of the first $(m - 1)$ be:

$$\text{const. } \theta_1^{x_1} \, \theta_2^{x_2} \ldots \theta_m^{x_m},$$

on their sample space,[1] where $\theta_m = 1 - \theta_1 - \ldots - \theta_{m-1}$, and each $x_i \geq 0$ is an integer. Let γ_i, $i = 1, \ldots, m$, be a set of constants,

$$V = \sum_{i=1}^{m} \gamma_i \Theta_i,$$

$n = \Sigma \, x_i$, $\bar{\gamma} = \Sigma \, \gamma_i x_i / n$. Then if $n \to \infty$ and each $x_i \to \infty$ in such a way that $r_i = x_i / n > 0$ and is kept constant, the random variable:

$$\sqrt{n} \, (V - \bar{\gamma}) / \sqrt{\Sigma \, (\gamma_i - \bar{\gamma})^2 x_i / n},$$

approaches the standard Normal $[\mathfrak{N}(0,1)]$ Distribution in law.

The statement of the theorem was generalized in a manner to be explained in the sequel, and a rigorous proof was given in 1919 by Mises (see Mises, 1964, pp. 352–357). Even with no further commentary a proof is seen to be a sophisticated technical achievement even by modern standards, let alone those of 1834, the time of presentation of Bienaymé's paper to the Academy.

The motivation for the theorem rests within the framework of the proposition, which, until recent times, was known as the "inverse" of the Moivre–Laplace Theorem, which, recall, refers to n Bernoulli trials, with probability p of success in each. Then, if P is the actual proportion of successes obtained, the Moivre–Laplace Theorem may be stated as:

$$Pr\left\{ p - t \sqrt{\frac{p(1 - p)}{n}} \leq P \right.$$

$$\left. \leq p + t \sqrt{\frac{p(1 - p)}{n}} \right\} \to (2\pi)^{-1/2} \int_{-t}^{t} e^{-x^2/2} \, dx,$$

as $n \to \infty$. Thus, if p is known and n is large, the theorem is useful for evaluating probabilities of the Binomial Distribution. We may also rewrite the left-hand side as:

$$Pr\left\{ P - t \sqrt{\frac{p(1 - p)}{n}} \leq p \leq P + t \sqrt{\frac{p(1 - p)}{n}} \right\},$$

[1]The corresponding probability distribution is now named after Dirichlet.

in which case the expression would be useful for obtaining a *confidence interval* for an unknown constant probability, p, of success if n is large, if the quantities on the extreme right and left did not themselves involve p. Nowadays, we tend to avoid this problem by substituting, in the two radicals, P for p, justifying this approximation by Bernoulli's Theorem and a theorem in Cramér (1946, p. 254), if n is large. We thus say, making a slight notational change, that:

If θ is the probability of success in each of a very large number n of Bernoulli trials in which x successes occur, then the probability that θ is contained in the random interval:

(5.1)
$$\frac{x}{n} \pm c \sqrt{\frac{x(n-x)}{n^3}},$$

is approximately

$$\sqrt{2/\pi} \int_0^c e^{-x^2/2} \, dx.$$

This proposition is known as the "inverse," mentioned before. Both Laplace and Poisson had felt the substitution of x/n for θ in the Moivre–Laplace Theorem to be valid and mathematically adequate; however, both were aware, as consequently was Bienaymé, of an earlier attempt by Bayes (1763) to put the inverse proposition on a more formal footing (see Czuber, 1899, p. 93), although this was not altogether successful. It is, in the first instance, with rigorous validation of the inverse proposition by a Bayesian approach, that Bienaymé (1838b) is concerned. Before proceeding, it is relevant to mention that the validity of the substitution step has been much debated, even as recently as by Gini (1955, pp. 334–336).

Bienaymé considers in place of θ the more general linear form:

$$v = \gamma_1 \theta + \gamma_2 (1 - \theta),$$

in which case its relevant estimate is:

$$\{\gamma_1 x + \gamma_2 (n - x)\}/n.$$

By noticing that the radical in (5.1) may be written as:

$$n^{-1} \sqrt{x\left(1 - \frac{x \times 1 + (n-x) \times 0}{n}\right)^2 + (n-x)\left(0 - \frac{x \times 1 + (n-x) \times 0}{n}\right)^2},$$

and that the case (5.1) corresponds to $\gamma_1 = 1$, $\gamma_2 = 0$, he deduces that in his case, in place of (5.1), the formula:

$$\frac{\{\gamma_1 x + \gamma_2 (n - x)\}}{n}$$

$$\pm \frac{c}{n} \sqrt{x \left(\gamma_1 - \frac{x\gamma_1 + (n - x)\gamma_2}{n}\right)^2 + (n - x)\left(\gamma_2 - \frac{x\gamma_1 + (n - x)\gamma_2}{n}\right)^2},$$

will presumably obtain. He credits Laplace (1812) for formulating the problem in this weighted form; his own ultimate intention is to establish a generalization beyond this, using tools provided by Laplace and, later, Poisson. After treating the above case, which may be described as the binomial, he goes on to the trinomial, and ultimately to the general multinomial. There is no basic difference in technique, so we shall sketch his reasoning in the multinomial setting.

Suppose at each of n independent trials, m possible mutually exclusive events can occur, with probabilities θ_i, $i = 1, \ldots, m$, respectively, and the number of times the ith type of event occurs is x_i. Thus:

$$1 = \sum_{i=1}^{m} \theta_i, \qquad \sum_{i=1}^{m} x_i = n.$$

Bienaymé's approach is Bayesian, for he treats the parameters θ_i, $i = 1, \ldots, m$, as random variables Θ_i, $i = 1, \ldots, m$. His basic problem, then, is to find the distribution of the random variable:

$$V = \sum_{i=1}^{m} \gamma_i \Theta_i \quad (\text{where } \gamma_1 \leq \gamma_2 \ldots \leq \gamma_m, \text{ w.l.o.g.})$$

conditional on the observed values x_i, $i = 1, \ldots, m$ (thus a *posterior* distribution); and then to investigate the effect of letting $n \to \infty$. He writes down the posterior distribution of the Θ_i, $i = 1, \ldots, m$, given the observations, in the form (commensurate with our mathematical statement at the beginning of this section):

$$(5.2) \qquad f(\boldsymbol{\theta}|\mathbf{x}) = \frac{\theta_1^{x_1} \theta_2^{x_2} \ldots \theta_{m-1}^{x_{m-1}} \theta_m^{x_m}}{\int \theta_1^{x_1} \theta_2^{x_2} \ldots \theta_{m-1}^{x_{m-1}} \theta_m^{x_m} d\boldsymbol{\theta}},$$

where $0 \leq \theta_i \leq 1$, $i = 1, \ldots, m$, $\sum_i \theta_i = 1$ defines the sample space. While he does not tell us what *prior* distribution he needs to achieve this, it is evident that (5.2) will result if the $(m - 1)$ random variables Θ_i, $i = 1, \ldots, m - 1$, are assumed to jointly have uniform density on their sample space.

The density of V may then be obtained by transformation and integration, say by:

$$(5.3) \qquad f(v|\mathbf{x}) = \frac{\int \ldots \int \theta_1^{x_1} \theta_2^{x_2} \ldots \theta_m^{x_m} d\theta_3 \ldots d\theta_m}{\int_v \{\int \ldots \int \theta_1^{x_1} \theta_2^{x_2} \ldots \theta_m^{x_m} d\theta_3 \ldots d\theta_m\} dv},$$

θ_1 and θ_2 each having been expressed as functions of v, $\theta_3, \ldots, \theta_m$, by virtue of:

$$\sum_{i=1}^{m} \gamma_i \theta_i = v, \qquad \sum_{i=1}^{m} \theta_i = 1.$$

However, following Laplace, Bienaymé prefers first to make a transformation, with a view to evaluating the integrals in (5.3) and then the integral:

$$\int_{a_1}^{a_2} f(v \mid \mathbf{x}) \, dv,$$

which will yield the probability associated with what we now call the Bayesian confidence interval for V, $a_1 \leqslant V \leqslant a_2$, which is essentially what is needed for the conclusion stated at the outset of this section.

The simple transformation, passing from variable V to variable U, is given by:

$$v = u + \bar{\gamma},$$

while the θ_i are transformed via $\theta_i = z_i + r_i$, $i = 1, \ldots, m$, using notation introduced earlier, so that:

$$\sum_{i=1}^{m} z_i = 0, \qquad \sum_{i=1}^{m} \gamma_i z_i = u.$$

Thus the numerator in the evaluation of $P(a_1 \leqslant V \leqslant a_2)$ is given by:

(5.4) $\quad r_1^{x_1} r_2^{x_2} \ldots r_m^{x_m} \displaystyle\int_{b_1}^{b_2} \int \ldots \int (1 + z_1 r_1^{-1})^{x_1} \ldots$

$$(1 + z_m r_m^{-1})^{x_m} dz_3 \ldots dz_m du,$$

where $b_i = a_i - \bar{\gamma}$ (if any x_i is 0, α^{x_i} is to be understood as unity for any α), while the denominator may be treated in like manner. Bienaymé writes:

$$\prod_{i=1}^{m} (1 + z_i r_i^{-1})^{x_i} = \exp\left\{ \sum_{i=1}^{m} x_i \log (1 + z_i r_i^{-1}) \right\},$$

so that the integrand in (5.4) becomes:

(5.5) $\quad \displaystyle\sum_{i=1}^{m} x_i \left(z_i r_i^{-1} - \frac{z_i^2 r_i^{-2}}{2} + \ldots \right) = -\sum_{i=1}^{m} x_i z_i^2 r_i^{-2} + S,$

(since $\sum x_i z_i r_i^{-1} = n \sum z_i = 0$), where S represents terms in higher powers in z_i. Proceeding hence in great detail as to the effect of letting $n \to \infty$, he finally concludes that, in the limit, V lies in the interval:

(5.6) $\quad \bar{\gamma} \pm cn^{-1} \left(\displaystyle\sum_{i=1}^{m} x_i (\gamma_i - \bar{\gamma})^2 \right)^{1/2},$

with probability $\sqrt{2/\pi} \displaystyle\int_{0}^{c} e^{-x^2/2} \, dx$, as we had stated at the outset.

The theorem was in fact later shown to hold for arbitrary joint bounded prior density for the random variables Θ_i, $i = 1, \ldots, m - 1$, by Mises as cited; he called this result the "second fundamental limit theorem," since it can be regarded as a partial converse to the "first fundamental limit theorem" which is the Central Limit Theorem. In a Bernoulli trials setting, Mises's result is given in the textbook of Bernstein (1946b), where it may already have occurred in an edition of 1917.

In connection with Bienaymé's working, Lancaster (1966, §3) has pointed out that the term in (5.5),

$$\sum_{i=1}^{m} x_i z_i^2 r_i^{-1} = \sum_{i=1}^{m} (x_i - n\theta_i)^2 / x_i - \sum_{i=1}^{m} (x_i - n\theta_i)^2 / n\theta_i,$$

is almost Pearson's Chi-square goodness-of-fit statistic.[2] However, it is unlikely that Bienaymé was thinking in this direction at the time, or that he *intended* to prove a Central Limit Theorem (analogous to the Moivre–Laplace Theorem for the Binomial Distribution) for the Multinomial Distribution, as Lancaster is inclined to believe, although it is evident that Bienaymé had the tools to do so.

The real significance of the Bienaymé–Mises result is that, asymptotically as $n \to \infty$, the specific assumption about the prior distribution becomes irrelevant, and the Bayesian confidence interval (5.6) for the linear combination $\Sigma \gamma_i \theta_i$ coincides with that obtained by the Central Limit Theorem for multinomial trials in conjunction with the Laplace substitution of x_i/n for θ_i. This illustrates again a point we have made elsewhere[3]: that for large samples, differences between approaches, whether of the English school (classical), nonparametric, or Bayesian, tend to disappear.

An interesting light on Bienaymé's result is provided by a point frequently made by mathematical statisticians of the Bayesian persuasion, such as Lindley (1965) and L. J. Savage: that use of uniform prior distribution yields a posterior density proportional to the likelihood function. The reader may already see this manifested in (5.2). Hence in finding the maximum likelihood estimator for the parameter $v = \Sigma \gamma_i \theta_i$ in the classical approach, we shall, under good conditions, be finding the mode of the posterior distribution of V in the Bayes approach (although the "Bayes estimator" is usually taken to be the mean). In any case, the maximum likelihood estimator of v is in fact $\bar{\gamma}$, and, under good conditions, such estimators are asymptotically Normally distributed, which ultimately gives us, once again, Bienaymé's conclusion.

There are two items of historical interest in connection with this paper of Bienaymé which are worth mentioning in conclusion to this discussion. The first is a report by a jury including Poisson on the paper in the year of its presentation to the Academy (Libri, 1834). A rather long affair, a large part

---

[2]See §3.4.
[3]See §3.4.

of it is concerned with trying to establish practical motivations and contexts for results of the kind, and pointing out the paramount importance in "the higher questions of social arithmetic" of probability calculus in general, and of probable error in particular. The second is a commentary on the paper by Merriman (1877) in characteristically direct, if often unperceptive, style, on the mathematical aspects:

> The rule of LAPLACE here meant is a method for finding the probability of the error of the mean. The opening pages contain some interesting historical remarks, but the investigation itself is very long and tedious and seems to be of little value.

5.3. Medical statistics

Bienaymé's abilities in the practical use of statistics are well evidenced by the comments in his paper *Application à la statistique médicale* (1840a) which was read to the *Société Philomatique* in February 1840. The paper concerns the use of statistical methodology in analyzing the effects of treatments on individuals. No mathematics is given and the setting is rather nonspecific; the advice given could equally be applied to treatments in various nonmedical contexts.

Bienaymé raises various issues. He points to the common error of using a simplistic probabilistic treatment in a complex case for which it has not been designed. Next, he observes that the real effect of a treatment under consideration is difficult to determine, since some individuals who are cured would have been cured without the treatment. There is a similar difficulty for differences between treatments. He notes, however, that this latter difficulty can be avoided:

> Fortunately, however, precise knowledge of this difference is not necessary to distinguish between two treatments used in identical situations, since the probability that one of the effects will surpass the other, without stating to what extent, becomes independent of the unknown divisor which was involved.

Here we presumably have the Central Limit Theorem applied to a (possibly paired) difference. Let $\bar{X}_i$, based on a sample of size n_i, be the mean of observations on treatment i ($i = 1,2$). Then, from the Central Limit Theorem, $\bar{X}_i$ is approximately distributed as $\mathfrak{N}(\mu_i, \sigma_i^2 n_i^{-1})$ for large n_i, where μ_i and σ_i^2 are the population mean and variance, respectively. Suppose $\mu_1 = \mu_2 + \delta$, some $\delta > 0$. Then, Bienaymé is presumably concerned with:

$$Pr(\bar{X}_1 < \bar{X}_2) = Pr\left(\frac{(\bar{X}_1 - \delta) - \bar{X}_2}{(\sigma_1^2 n_1^{-1} + \sigma_2^2 n_2^{-1})^{1/2}} < \frac{-\delta}{(\sigma_1^2 n_1^{-1} + \sigma_2^2 n_2^{-1})^{1/2}}\right) \to 0,$$

as n_1 and n_2 approach infinity, whatever be δ, σ_1^2, σ_2^2. The principal application in mind involves the Bernoulli case where the possibilities are death or recovery.

Bienaymé then proceeds to comment on the matter of sample size, showing a reasonable understanding of ideas underlying the testing of statistical hypotheses via the above mathematics and a good practical sense:

> Only then, for the comparison to have any validity, will it be necessary to ensure that the number of observations is sufficiently large. The probability obtained may be very small, either because one of the treatments is in reality only very slightly preferable to the other, or because both will have been applied to too small a number of patients in danger of not recovering, for the difference to be able to acquire any certainty. This indecision will necessarily occur in many cases in which common sense alone cannot judge. It will be possible to make a successful distinction only by estimating approximately the number of patients who would have been able to recover without the treatments which are compared.

The last sentence suggests a forerunner of the t test, namely the replacement of the variance by its sample estimate, although this is not notable for the time. Others, and in particular Laplace, had previously used similar ideas. The t test itself was introduced by W. S. Gosset, writing under the pseudonym Student (1908).

5.4. The Law of Averages

In his paper (1840b), Bienaymé remarks on common misinterpretations of what would now be known as the "Law of Averages." Discussion of these misinterpretations was not new even at this time and Bienaymé states that:

> Popular works on probability point out this error.

However, the explanations which had been given were regarded as superficial by Bienaymé.

There are two illustrations discussed in the paper, the first concerning monthly rainfall data in the case where the mean annual rainfall varies little from year to year. Here the error concerns the anticipated rainfall in the last 9 months of a year in which the first 3 months produce a total rainfall near to the usual annual total. A description is given of how to formulate appropriate conditional probability statements in order to correctly interpret such phenomena:

> In fact, the high probability of the almost invariable mean annual result supposes essentially that the distribution of rainfall over the whole year can be arbitrary. There is therefore no comparison to be made between this probability and that

which must relate to a special rainfall distribution. In this specific case the compound event, for which the probability is to be calculated, consists of a heavy rainfall for three months, followed by a severe drought. And this probability, whatever it may be, will have to be divided by that of a heavy rainfall for three months calculated in isolation. These two unknown probabilities are not formed directly in the summaries of metereological observations, but they can be deduced from them. Only then will it be known whether there is any reason to expect a coming drought when the mean quantity of the annual rainfall has been almost reached or, equally, if even further rains may be expected.

Bienaymé makes no suggestion that the monthly data may be independent. His reasoning is soundly based on probabilities estimated from the meteorological data.

The other example which Bienaymé discusses concerns the absence of a particular number in a long sequence of lottery results. Here the common error lies in implicitly building a memory into a sequence of independent trials. Bienaymé again goes through the full conditional probability argument and observes that:

. . . past results could not have any influence on future results. . . .

The arguments given are mathematically very simple. The phenomena under consideration are divided into events past P and future F. Any predictions based on a knowledge of P should be derived from:

$$Pr(F|P) = Pr(P \cap F)/Pr(P).$$

Of course,

$$Pr(F|P) = Pr(F),$$

if P and F are independent.

The kind of error to which Bienaymé is addressing his comments is, of course, still endemic. A rather typical illustration is given by the following argument of Bradman (1950, p. 209):

Once again my call of heads proved wrong and my career in England as Australia's captain ended by my losing my eighth toss out of nine in Test Matches, calling heads every time. Surely this must be a warning for my successor to call heads. The law of averages did not work in my case but in the long run it must.

The motivation for Bienaymé's paper appears to be a defence of the lottery which had been run by the state in replacement of the tontines.[4] He writes:

. . . the paralogism . . . is not explained clearly: and this is undoubtedly why it has persisted strongly during the fifty years of the existence of the lottery. Each time the absence of a number is noted in a series of draws of any length, the receipts of the public funds are increased.

---

[4]See §2.3.

5.5. Electoral representation

The paper (Bienaymé, 1840c) which was read to the *Société Philoma-tique* in March 1840 is concerned with the following problem and its electoral system application. A large number of packets of cards are taken at random from a very large number of cards of two colors which are in known proportion. The problem is to determine the number of packets which have a majority of cards of one color, specified in advance. The number of packets taken may exhaust the supply of cards.

The electoral system application involves the case of a very large number of voters divided between two opinions in a known ratio. Voters are distributed at random into a large number of electorates, and the problem is to predict the number of electorates in which the majority hold a specified opinion.

Bienaymé states that no rigorous solution to this problem has existed prior to his work. In fact, he claims that previous work has made invalid application of a correct proposition, namely that the probability of finding a majority of cards of the specified color in one of the packets chosen at random depends only on the number of cards in the packet. No attribution is given in the paper, but from the bibliographic note, Bienaymé (1852b), we learn that the previous work in question was that of Poisson (1837b). The report of Bienaymé's contribution states:

> . . . that he had noticed the defect as soon as it was published. This defect, in fact, exists in a quite similar form in the solution of the game of *trente et quarante*, in which M. Bienaymé had previously recognized it.

In connection with the game of *trente et quarante,* Bienaymé is presumably referring to the work of Poisson (1820) which had aroused sufficient interest to be published three times.[5] We note that Poisson died on 25 April, 1840, approximately a month after Bienaymé's paper was read. It seems likely that Bienaymé did not mention his name explicitly out of deference to the circumstance of Poisson's illness.

The formal details of Bienaymé's solution are, as usual, not provided; only the following intriguing comment is reported:

> . . . its solution may, from the logical point of view, be of some interest, since it is by means of a pure analytical artifice that the problem can successfully be separated from a group of arguments which are very difficult to follow. . . .

Perhaps he is referring to an application of generating functions. He does, however, give an example. If there are 208,000 voters distributed into 440 electorates and one opinion has a majority of one-twentieth of the voters, then approximately 85% of the electorates will have the majority opinion.

---

[5]Nevertheless, the literature on the game extends back to Florence (1739).

He also remarks on the phenomenon that has resulted in a small majority of 52.5% of voters leading to victory in a large majority of 85% of electorates:

> From this disproportion between the ratio of the numbers of colleges and the ratio of the majority to the minority, a consequence which is unfavorable to the system of elections has been inferred. It has been noted that an opinion which has a very strong minority would have only a small number of representatives, and that thus the representative system would be merely a deception, without the influences created by the meeting of electors of the same locality and other similar causes.

Bienaymé's numerical results can be approximately checked very quickly using the Normal approximation to the Binomial and this, in fact, would appear to be the invalid solution to which he is referring. Bienaymé notes that such a solution does not take account of the influence of the composition of various packets on one another but he remarks that the correction will provide little change to the numerical answer. This is explained as follows:

> M. Bienaymé recalls that the probabilities of results concerning large numbers are normally expressed by an integral . . . the limits of which themselves decide limits which must be attributed to the most probable values. Moreover, these latter limits are composed of two terms, one of which is proportional to the number of events considered, whereas the second term is only proportional to the square root of this number. Now, in the present question, it happens that the term proportional to the number of packs of cards, or electoral colleges, was not affected in any way by the inaccuracy of the argument. This affected only the second term, which determines the magnitude of the limits of the probable values, and which is only proportional to the square root of the number of colleges. The rigorous solution will thus modify only the extent of these limits.

It seems possible that Bienaymé's solution involved the Multivariate Normal approximation to the Multivariate Hypergeometric Distribution. Indeed, suppose we have N electorates with n_i voters in the ith, $1 \le i \le N$, and denote the number of votes in favor of a specified opinion in the ith electorate by S_i. The exact probability in question is derived from the Multivariate Hypergeometric; for example, the probability that *every* electorate is in favor is:

$$Pr\left(\bigcap_{i=1}^{N} \{ S_i > [n_i/2] \} \;\Big|\; \sum_{i=1}^{N} S_i = S = \alpha \sum_{i=1}^{N} n_i \right)$$

$$= \sum_{\substack{x_1 > [n_1/2] \\ x_1 + \ldots + x_N = S}} \cdots \sum_{x_N > [n_N/2]} \frac{\prod_{j=1}^{N} \binom{n_j}{x_j}}{\binom{n_1 + \ldots + n_N}{x_1 + \ldots + x_N}} \Bigg/ \sum_{x_1 +} \cdots \sum_{x_N = S} \frac{\prod_{j=1}^{N} \binom{n_j}{x_j}}{\binom{n_1 + \ldots + n_N}{x_1 + \ldots + x_N}}$$

where α ($>1/2$) is known and $[x]$ denotes the integer part of x. Bienaymé certainly had a complete understanding of the properties of the Hypergeometric Distribution and the use of asymptotic Normality in this context. Such issues were discussed in considerable mathematical detail in his paper (1840d), read to the *Société Philomatique* in April 1840, just one month after the present paper. The results of this later paper, which is one of Bienaymé's more significant works, are discussed in the next section and have also been mentioned in §3.5.

At least from the time of Condorcet there had been much interest in probabilistic questions related to electoral systems, so in a sense the work of Bienaymé was a product of the period. Bienaymé did, however, stimulate further consideration of this particular problem. For example, Bertrand (1888a) begins as follows:

> It has been noted several times that on dividing a country up into electoral colleges in which chance would associate a specific number of electors, the result would be far from representative of the proportion of opinions which divide the country. The dominant opinion would be favored to an extent which is revealed by accurate calculation and which exceeds any prediction.

Bertrand provides further elucidation of the issue raised by Bienaymé, although without mention of Bienaymé, and via a rather different approach.

Bienaymé's work, at least through its likely motivation of the work of Bertrand, is linked to the development of what is now known as the classical ballot problem. This deals with two candidates in an election and the probability of one candidate always leading the other in counting, conditional on the final allocation of votes being known. For a historical discussion of the ballot problem see Takács (1967, Chapter 1). The classical result was found by Bertrand (1887) and important contributions by Barbier and André followed in the same year. Takács traces the antecedants back to work of Moivre (1711) in 1708 and includes Lagrange (1775), Ampère (1802), and Laplace (1812), but not Bienaymé.

5.6. The concept of sufficiency

In his 1840d paper Bienaymé chiefly concerns himself with a discussion of the stability of relative frequencies, particularly with respect to disjoint partial series. Motivation was provided by work of Fourier (1821, 1826, 1829) about which Bienaymé writes:

> Fourier, in the *Recherches statistiques sur Paris*, had advised separating the observations into groups in order to ascertain from the deviations of the partial

results whether it is possible to place any reliance on the general mean. He did not, however, give any rule on this matter. Uncertainty therefore persisted.

The issue of stability of relative frequencies is treated in §3.5. However, from a modern point of view the paper is most interesting and important as the results show that Bienaymé understood the arguments leading to the concept of *sufficiency*, at least in the context of the Multinomial Distribution. He starts with the Binomial. Let X_1 and X_2 be independent, and have Binomial Distributions:

$$Pr(X_1 = k) = b(k; m,p) = \binom{m}{k} p^k(1 - p)^{m-k}, \qquad 0 \leq k \leq m,$$

$$Pr(X_2 = k) = b(k; n,p) = \binom{n}{k} p^k(1 - p)^{n-k}, \qquad 0 \leq k \leq n.$$

Bienaymé calculates:

$$Pr(X_1 = r, X_2 = a - r | X_1 + X_2 = a) = \binom{m}{r}\binom{n}{a - r} \bigg/ \binom{m + n}{a},$$

$$0 \leq r \leq \min(m,a),$$

and observes that it is independent of p. He notes, furthermore, that this property is possessed by a general decomposition into component parts:

> The probability relationships between the partial series and the total series of tests or experiments are therefore not only independent of the actual probability of the events, but also they are the same as if the results of which a partial series is composed had been selected at random from the entire series of results observed.

He then makes the following comment which indicates his grasp of the idea for the Multinomial case:

> The application of this principle (which also extends to all cases of constant probabilities, whatever the number of types of events of which the result is composed) will be very simple.

Bienaymé did not advocate the use of a sufficient statistic. He is principally concerned, as we have noted in §3.5, with testing for homogeneity, and the real message of his paper, as he saw it, is described as follows:

> This is an important consequence, since from this it emerges that statistics, and observational sciences in general, can always provide positive data on the constancy of natural laws, independently of the form taken by these laws.

Nevertheless, the concept of sufficiency is essentially at hand, albeit not explicitly recognized. The factorization:

$$Pr(X_1 = r, X_2 = a - r) = P(X_1 = r, X_2 = a - r | X_1 + X_2 = a)P(X_1 + X_2 = a)$$
$$= \left[\binom{m}{r} \binom{n}{a-r} \Big/ \binom{m+n}{a} \right]$$
$$\times \left[\binom{m+n}{a} p^a (1-p)^{m+n-a} \right],$$

indicates so clearly that the grand mean provides all the information that the observations have to offer about p that it would be rather surprising if this implication were lost on Bienaymé. He is, after all, specifically considering the grand mean and its relation to the subsample means and writes, of homogeneous trials:

> One can show without difficulty that, in this case, the probability relationships which must exist between the general result [i.e., the grand mean] and the partial results are absolutely independent of the probability of the phenomena. . . .

The discovery of concept of sufficiency is currently attributed to R. A. Fisher (1920) and it is interesting to contrast the genesis of his ideas on the subject with those of Bienaymé. Fisher began by assuming that each of n observations is Normally distributed with mean m and variance σ^2 and the object was to estimate σ. He focused attention on the estimator σ_1, the mean deviation from the sample mean, and σ_2, the sample standard deviation. He derived the variances of both σ_1 and σ_2, and observed that for large n a comparison of these should be adequate for discrimination since both are approximately Normally distributed. However, for small n the situation is more complicated, and to clarify this situation Fisher derived the joint distribution of σ_1 and σ_2 for the case $n = 4$. This enabled him to observe that (the italics are Fisher's):

> *For a given value of σ_2, the distribution of σ_1 is independent of σ.* On the other hand, it is clear . . . that for a given value of σ_1, the distribution of σ_2 does involve σ. In other words, if, in seeking information as to the value of σ, we first determine σ_1, we can still further improve our estimate by determining σ_2; but if we had first determined σ_2, the frequency curve for σ_1 being entirely independent of σ, the actual value of σ_1 can give us no further information as to the value of σ. The whole of the information to be obtained from σ_1 is included in that supplied by a knowledge of σ_2.
>
> This remarkable property of σ_2, as the methods which we have used to determine the frequency surface demonstrate, follows from the distribution of frequency density in concentric spheres over each of which σ_2 is constant. It therefore holds equally if σ_3 or any other derivative be substituted for σ_1. If this is so then it must be admitted that: *The whole of the information respecting σ, which a sample provides, is summed up in the value of σ_2.*

The vital difference between this work and that of Bienaymé is that Fisher was focusing on the question of estimation, whereas Bienaymé was not. Cournot (1843, footnote, pp. 201–203), however, does interpret Bienaymé's work in an estimation context but only in a superficial fashion.[6]

The term "sufficiency" is not introduced in Fisher's (1920) paper, nor is the idea explored for other underlying distributions. However, the significance of the concept was not lost on Fisher, and in Fisher (1922) he names sufficiency, relates it to maximum likelihood, states the Factorization Theorem, and applies the concept to a variety of situations. This material is further refined in Fisher (1925). See Stigler (1973) for a more detailed discussion of Fisher's work and motivation and for comments on much earlier related work of Laplace (1818), who was some distance from isolating the concept of sufficiency and was concerned with a very different problem from that of Bienaymé.

Laplace's problem was one of estimation of y in the regression model

$$p_i y - a_i + x_i = 0, \qquad i = 1, 2, \ldots, n,$$

where the p_i's and a_i's are known, and y and the x_i's are unknown, the x_i's being errors of observation. Laplace considered two estimators, one the least-squares estimator, and one that might now be called a weighted median. He went on to obtain the joint distribution of the difference between y and each of these estimators and from this deduced that least squares provided a better (in the sense of asymptotic variance) estimator of y than the weighted median and, furthermore, is also better than a linear combination of the two. There is no evidence, however, of any ideas relating to conditional distributions or of information content of the sample in the estimators.

5.7. A general inequality

Let $a = \{a_i\}$ and $\{c_i\}$ be sets of positive numbers. In the 1840e paper, Bienaymé announces the result that:

$$M_m(a) = \left(\sum_{i=1}^n c_i a_i^m \bigg/ \sum_{i=1}^n c_i \right)^{1/m},$$

is monotonic nondecreasing in $m \geq 0$. Writing this as:

(5.7) $M_r(a) < M_s(a), 0 \leq r \leq s,$

unless all the a_i are equal, we see that the case $s = 2r$ is Cauchy's (1821) Inequality, which was well known and alluded to by Bienaymé, while the

---

case $r = 0$, $s = 1$ is the Arithmetic–Geometric Mean Inequality. Bienaymé deduces the inequality:

$$n^{-1}(a_1 + \ldots + a_n) < \left(a_1^{a_1} \ldots a_n^{a_n} \right)^{(a_1 + \ldots + a_n)^{-1}},$$

possibly by taking the case $r = 0$, $s = 1$ with the a_i replaced by a_i^{-1} and $c_i = a_i$. This complements the simple form of the Arithmetic–Geometric Mean Inequality:

$$n^{-1}(a_1 + \ldots + a_n) > (a_1 \ldots a_n)^{n^{-1}}.$$

Bienaymé cites no references and gives no proofs. He merely prefaces his principal result by citing what is now known as the Cauchy Inequality and then goes on to say that:

> It is then very easily seen that this proposition is only a particular case of another more general one . . .

The case where all the c_i are the same had in fact been established earlier by Reynaud and Duhamel (1823) (quoted in Chrystal, 1952, p. 49). Bienaymé's work escaped attention and his general result is attributed to Schlömilch (1858) in the authoritative source Hardy, Littlewood, and Polya (1967, p. 26).

Curiously, the result (5.7) is proved later, with no citation, by Liapounov (1901, §1) and is instrumental in the proof of that general version of the Central Limit Theorem, now known as Liapounov's Theorem, which is the subject of his paper. Indeed, the general inequality of which (5.7) is a manifestation, viz.,

$$E^{1/r}\{|X|^r\} \leq E^{1/s}\{|X|^s\}, \qquad 0 \leq r \leq s,$$

has come to be known as Liapounov's Inequality in the probabilistic literature. It is, further, interesting that Liapounov's Theorem (with remainder term) can be used to deduce the ultimate Lindeberg–Feller version of the Central Limit Theorem, as demonstrated by Bernstein (1946c). Thus, in a sense, through result (5.7) Bienaymé also occupies a niche in the history of the Central Limit problem[7] and the "first fundamental limit theorem," quite apart from his contribution to the "second fundamental limit theorem" as discussed in our §5.2.

5.8. A historical note on Pascal

The associations of Blaise Pascal with the foundations of probability theory are well known and concisely described by Ore (1960). The note of Bienaymé (1843), pertaining to them, is apparently totally unknown, even

---

[7]See §5.10 for further support for this claim.

though it has what appear to be a number of new observations to make; it is unfortunate that Bienaymé did not list it even in his own bibliography (Bienaymé, 1852b) of scientific writings. It is apparent, however, from a concluding comment:

> . . . this very citation by Bernoulli assures France of total priority in the invention of the probability calculus.

that Bienaymé's motivation suffered from a nationalistic spirit of inquiry which is frequent (and unfortunate) even among today's chroniclers of the history of probability, and led to a not entirely correct conclusion.

The first observation concerns the famous passage: *De la nécéssité du pari* (Pascal, 1958, Article III) or *The Wager* (Pascal, 1966, Article II of Section 2), from the *Pensées*. This is an argument, aimed at the agnostic (not the atheist), with a view to forcing him to live, to all intents and purposes, in conformity with Christian morality (Steinmann, 1965), which we now recognize as decision–theoretic. Pascal argues that the rational individual will base a decision on what is in fact a mathematical expectation of gain (that is, on the prior risk); and if an individual permits even a small prior probability of God's existence, the expected gain is infinite. Pascal's argument also has in it elements of the idea of modifying the prior distribution on the basis of further evidence, such as, in this case, the Scriptures. This is all well described by Ore, whose article, however, gives the impression that he is the first to notice that the argument turns on the notion of mathematical expectation; however, that this is so is already known to Bienaymé. It is, nevertheless, perhaps a little strong to say (cf. Hacking, 1975) that Pascal invented decision theory.

To understand properly the other points of Bienaymé's morsel, it is unavoidable to give a résumé of Pascal's life and work. Our investigations here are based on the recent excellent general account of the Jesuit scholar Steinmann (1965), and on Pascal's annotated *Oeuvres*. Of course Pascal's productivity extends far beyond probability and mathematics, yet elements of these fields occur in his theological and philosophical writings. It is therefore not unlikely that, like other scholars, we have still missed relevant information, for it is largely with a topic of this nature that the sequel *vis-à-vis* Bienaymé is concerned.

Pascal was born in 1623 and died in 1662. The Jansenist controversy had its origins with the publication in 1640 by Jansen (Jansenius) of the book *Augustinus,* concerned with the doctrine of grace, going back to St. Augustine and beyond the commonly received teaching on grace at the time. The Jansenist influence, distinguished by its piety, was communicated to the religious community of Port-Royal, renowned, at a time of general religious laxity, for the fervor and strictness of its inmates. This community consisted of a convent, *Port-Royal de Paris,* governed by Mother Angélique Arnauld, and a band of sympathetic lay disciples—at the former rural residence of the nuns, *Port-Royal des Champs*—through which its influence on French thought was penetrating. Out of this community came a book by Dr. Antoine Arnauld concerned with frequent communion and five issues extracted by the Church from the *Augustinus,* which were the grounds on which the Church and the Jansenists were ultimately to rejoin battle in about 1660. The issues were heightened by the fact that the French Jansenists refused to submit to, or to openly leave, the Church; and the intervention of the Jesuit order, who became the Jansenists' archfoes, and against whom Pascal's brilliant *Provincial Letters* were eventually directed.

Pascal was first influenced by knowledge of the Port-Royal movement in 1645 at the age of 22, and one of his sisters, Jacqueline, entered its Paris house in 1652. The night of 23–24 November, 1654, saw his own famous (second) "conversion," to the same thinking; and shortly after he joined the lay community associated with the Granges at the rural convent of *Port-Royal des Champs.* He became interested in pedagogic research centered on the associated "little schools." For the older pupils among these children, Arnauld, with the assistance of Nicole and Pascal, wrote a "Logic." The first edition of 1662, which was anonymous and entitled *La Logique ou l'Art de Penser,* is now attributed, in a recent reprinting, to Arnauld and Nicole (1970) as authors; it may be useful for the reader to note that the work is also often referred to as the *Logique de Port-Royal.* Actually, Pascal's part in it remains obscure. We also need mention in this connection such comments as Steinmann's (1965, p. 182)[8]:

> . . . during the years of gestation of the *Pensées* Pascal . . . wrote a few pages of a treatise on logic. . . . They have been preserved under the title of *De l'Esprit géométrique* and *De l'Art de persuader.* . . . It is part of Pascal's *Discours de la méthode.*

Bienaymé's second point refers to the *Logique* in connection with Jacob (James) Bernoulli, specifically as a source of inspiration of Bernoulli's Theorem:

> One reads on p. 225 of the fourth part of his *Ars conjectandi* that his ideas have been suggested to him, partially at least, by Chapter 12 and the chapters following it, of *l'Art de penser,* whose author he calls *magni acuminis et ingenii vir* [a man of great acumen and ingenuity]. Now, this *Art de penser* is just the *Logique* of Port-Royal, published in the year of Pascal's death (1662). The final chapters contain in fact elements of the calculus of probabilities, applied to history, to medicine, to miracles, to literary criticism, to incidents in life, etc.; and are concluded by the argument of Pascal on eternal life.

In the 1970 reprint of the *Logique,* the fourth part is entitled *De la méthode* and contains the chapters which inspired Bernoulli. The reference "the argument of Pascal . . ." is to *The Wager.* (See Uspensky, 1937, pp. 105–106 for an English version of Bernoulli's comments.)

---

[8]The "years of gestation" are prior to and *circa* 1658.

It is now our task to elucidate as far as possible the question of Pascal's authorship and role in this setting. The two fragments mentioned by Steinmann are reprinted on pp. 240–290 of **9** of Pascal's *Oeuvres* (1904–1925). According to the editorial comment, they are presumed written in 1658–1659, and there is little doubt that they were sketches intended for the *Logique*. Indeed, an initial chapter of it entitled *Discours sur le dessein de cette Logique,* in a paragraph beginning on p. 15, refers to *De l'esprit géométrique* by name, which is said to have been:

> . . . a small unpublished manuscript, produced by a first-rate mind. . . .

In the same paragraph, it is admitted that not all the new reflections contained in the book are due to the authors of the book, and indeed in particular that *De l'esprit géométrique* greatly influenced Chapter X of the first part of the book, whose themes are further developed in the fourth part, which, however, contains, it is said, many extensions not in the influencing manuscript. The fourth part of the book treats the mathematical method, and it is now a traditional contention that Pascal wrote it, the comment just mentioned apart. Also opposing the contention is evidence in the *Notes* by Racine obtained in conversations with Nicole, that only he and Arnauld had worked on the book, and that the fourth part was the work solely of Arnauld.

It must be mentioned that Arnauld was a mathematician of some note in his time, and relations with Pascal, an indubitably greater mind, were less than cordial. It is also well to remember that Pascal actually completed very little of what he began; it is said (*Oeuvres* **3**, p. 297):

> As is the case with every creative mind, Pascal did not do what he had announced. The bulk of the treatises he had promised the Academy never saw the light of day. It is only with difficulty that we can discover any trace of them.

It is not clear whether the *Logique* appeared before or after Pascal's death. It is, however, known that Pascal was most concerned that due credit be always accorded to him; and his name is actually mentioned in the second edition.

We should mention for completeness that the two fragments of Pascal have been utilized also in several editions of his *Pensées,* namely those of Condorcet in 1776, and Bossut in 1779.

Bienaymé's last point is expressed in the final phrase:

> . . . the treatise which Pascal had edited and presented to the Academy of Science under the title *Aleae geometria* (Vol. IV, p. 410) seems lost forever.

Keeping in mind that "geometry" even in Bienaymé's time was used to refer to mathematics in general, our interest here is aroused by this apparently missing early treatise on probability. Actually, while the comment is characteristically obscure, it is uncharacteristically inaccurate, of Bienaymé. The facts appear as follows. We find that in 1654 (the year in which most of Pascal's contributions to probability were made), Pascal made an address to *L'Académie Parisienne* (*Academia Parisiensis;* at this time there were a number of unofficial "Academies," and this one seems identical with the *Académie de Montmor*, the unofficial predecessor of the later *Académie des Sciences.*) This address is recorded in the *Oeuvres* (**3**, p. 305–308) in Latin, reprinted from Bossut's Vol. IV, p. 408 *et seq.*, which is evidently the source Bienaymé is citing. In the address Pascal describes a number of projects on which he is working. Only two came to full fruition, as the commentary preceding the item in the *Oeuvres* remarks, one of which, on games of chance, came eventually to bear not the original name: *Aleae geometria*, since its contents were transformed into *Traité du Triangle Arithmétique*, thought to have been written *later* in 1654 and finally published posthumously in 1665, having only been discovered after Pascal's death (see Pascal, 1665; and *Oeuvres* **3**, 445 *et seq.*). This *Traité* treats extensively the formulation and solution of the division problem *(méthode de partis)*, on which the correspondence between Fermat and Pascal occurred in July to October 1654. Hara (1964), however, notes that Pascal's address to the *Académie* contains no mention of *this* problem, which suggests that the address was, indeed, *early* in 1654 and tends in small measure to support the implication of Bienaymé's statement that the *Aleae geometria* differed from the *Traité* substantially. On the other hand, Bienaymé makes no mention of the *Traité*.

It is, finally, interesting that p. 499 of the *Oeuvres*, **3**, within the context of a reprinting of the *Traité* carries a footnote which indicates that Johannes Bernoulli was well versed in Pascal's *Traité*.

* * *

While discussing Bienaymé's interests in the history of mathematics, we should perhaps devote a little attention to his note (Bienaymé, 1870) already mentioned in Chapter 1. In this he makes clear from two obscure primary sources in classical Greek that the Pythagoreans understood the concept of a *series*, beyond that of a geometric progression. The second, and older, of the two passages, is particularly obscure; but Bienaymé manages an apparently adequate explanation in terms of geometric progressions, each beginning with one, and with common ratio 3 and 2, respectively, displaying a remarkable perception.

5.9. The simple branching process

Until recently, it was thought that the theory of branching processes began with the work of Galton (1873) and Galton and Watson (1874) and was stimulated by the work of Candolle (1873), who followed on from

Doubleday (1842). Although many earlier authors had commented on the extinction of large numbers of noble families, Candolle pointed to the possibility of a probabilistic interpretation for this phenomenon. This evoked considerable interest as an alternative to earlier hypotheses, such as that of diminished fertility (as propounded by Doubleday).

Galton formulated the problem of extinction of families as follows:

> Let p_0, p_1, p_2, . . . be the respective probabilities that a man has 0, 1, 2, . . . sons, let each son have the same probability for sons of his own and so on. What is the probability that the male line is extinct after r generations . . . ?

Watson's solution using generating functions, however, contains an algebraic oversight and he incorrectly concludes that each family will eventually die out with probability 1. It was not until considerably later that this error was rectified, although its implications were strongly doubted by various authors. For example, Fahlbeck (1898, 1902), in a large work on the Swedish aristocracy, thought that he had refuted Galton and Watson, although not on mathematical grounds; he writes (Vol. 1, p. 133):

> There does not exist any inner necessity that the family must die out.

Haldane (1927), in applying the model in genetics, obtained what is tantamount to a completely correct statement of what is now known as the Criticality Theorem, while Steffensen (1930) gave for the first time a clear statement and detailed proof. Steffensen's solution was in response to the problem, posed afresh by A. K. Erlang (1929), who did not know of the earlier British work. What emerged (the Criticality Theorem) was that, writing:

$$f(s) = \sum_{j=0}^{\infty} p_j s^j, \qquad 0 \leq s \leq 1,$$

if $m = f'(1) \leq 1$ there is ultimate extinction with probability 1, while if $m > 1$ the extinction probability is the unique nonnegative solution less than 1 of the equation:

$$s = f(s).$$

For accounts of the historical development of the subject during this period see D. G. Kendall (1966), Albertsen (1973), and Jagers (1975).

The strikingly original earlier work of Bienaymé (1845) on this subject was first noted by Heyde and Seneta (1972). In his communication of 1845, anticipating Galton and Watson by some 28 years, Bienaymé shows that the correct statement of the Criticality Theorem is known to him:

> If . . . the mean of the number of male children who replace the number of males of the preceding generation were less than unity, it would be easily realized that families are dying out due to the disappearance of the members of which they are

composed. However, the analysis shows further that when this mean is equal to unity families tend to disappear, although less rapidly. . . .

The analysis also shows clearly that if the mean ratio is greater than unity, the probability of the extinction of families with the passing of time no longer reduces to certainty. It only approaches a finite limit, which is fairly simple to calculate and which has the singular characteristic of being given by one of the roots of the equation (in which the number of generations is made infinite) which is not relevant to the question when the mean ratio is less than unity.

This last paragraph presumably refers to what we would now write in the form:

$$q_{n+1} = f(q_n),$$

where q_n denotes the probability of extinction after n generations. Nevertheless, we are rather left in the dark as to Bienaymé's reasoning. As in the case with most of his *Société Philomatique* papers, no mathematical details are given and an interesting attempt to reconstruct his argument has been made by D. G. Kendall (1975). Bienaymé's paper ends with the tantalizing comment:

M. Bienaymé develops various other notions which the elements of the question have suggested to him, and which he proposes to publish in the near future in a special paper.

It is not at present clear, however, whether any such "special paper" was ever published. Certainly Bienaymé did not list it among his publications in his bibliography (1852b), published at about the time of his election to the Academy of Sciences.

Bienaymé gives no references in his 1845 paper but the title and the opening paragraph indicate that his motivation is similar to that of Galton:

A great deal of consideration has been given to the possible multiplication of the numbers of mankind; and recently various very curious observations have been published on the fate which allegedly hangs over the aristocracy and middle classes; the families of famous men, etc. This fate, it is alleged, will inevitably bring about the disappearance of the so-called *familles fermées*.

It seems likely that Bienaymé's work was stimulated by a paper of L. F. Benoiston de Châteauneuf entitled *Sur la durée des familles nobles de France*, first published in 1845. Châteauneuf, a demographer and statistician of note who had been working in this general area for some years, had already published several related papers: *De la durée de la vie chez le riche et chez le pauvre* (1830), and *De la durée de la vie chez les savants et les gens de lettre* (1841). The paper of Châteauneuf (1845) was read in two sessions, on 31 August, 1844 and 16 February, 1845, while Bienaymé's was read on 29 March, 1845. Châteauneuf's paper provoked lively discussion and the work certainly merits the description of " . . . observations . . . on the fate which hangs over the aristocracy . . .'' but his was not the only such work. In fact his paper begins in the following way:

> My esteemed colleague and friend, M. Villermé (1843), when giving an account to the Academy last year of a work by M. Doubleday on population (1842), expressed himself in this way: "The author deals in this work with the duration of aristocratic families, which, he assures us, is by no means as long as is thought. . . ."

Both Châteauneuf and Doubleday (who had an earlier and briefer account of his work published in 1837) discuss the average lifespan of a family name. Châteauneuf is the more concrete in his estimation, and writes:

> It was interesting to attempt to ascertain what the average lifespan of our historic houses could have been, the most ancient of which date back to the twelfth, eleventh, and tenth centuries and which have today ceased to exist, either completely or only in their oldest branch. In other words, after what average period of time did it happen, in the natural course of events, that the last descendant of a family died without living male issue, or leaving only daughters? The name was then *naturally* extinct. I found that this period of time, for 320 families, had been 300 years.

Later he refers to the work of Doubleday, following Malthus:

> In Switzerland . . . where the designation bourgeois is quite honorable, and important enough for many families to seek it, and where the names of those who achieve it are carefully recorded in a public register, according to M. Doubleday, or rather according to Malthus whose studies he quotes: of 487 families accepted, from 1583 to 1654, by the governing council into the Bern bourgeoisie, there remained, after a century, only half, and only 168 still existed in 1783. Two thirds had disappeared. Finally, of 112 families which made up the sovereign council of Bern canton, only 58, or half, still existed in 1796. M. Doubleday assures us that the same is the case for the rich and privileged middle classes in some cities in England, such as Newcastle-on-Tyne, Berwick-on-Tweed, etc.

Doubleday was concerned with drawing a distinction between the upper classes, whose names became extinct, and the others whom he supposed would explode in numbers. On the other hand, he had found some interesting early references on the subject and quotes Tacitus and, much more recently, Sir Thomas Browne (1658):

> Generations passe while some trees stand, and old Families last not three oaks.

Kendall (1975) suggests that we could count one oak (life to maturity) as 100 years, thus putting Browne in basic agreement with Châteauneuf.

We now return to Bienaymé's paper in which there are a number of other interesting aspects. He comments that the branching process does not allow a population to remain in a stationary state, in contrast to what authors of life tables suppose in their calculation. He makes gratuitous criticism of Poisson's Law of Large Numbers (see § 3.3), which does not apply in the present context. In addition, he also makes the following statement:

> . . . one is led to a very remarkable conclusion: this is that the ratio of generations could not have a permanent value, and that it must sometimes exceed unity and sometimes fall below.

Presumably the description "very remarkable" is attached because of the contrast with the *deterministic* exponential growth-rate models of Malthus and others. Bienaymé does not explicitly take up the question of population explosion in the supercritical case (mean number of offspring greater than unity), but he has noted the behavior of the means of the successive generations, which do grow exponentially, and would have been aware of the implications. Perhaps this material was intended for the "special paper." It was, after all, not very long since the fundamental idea that a population, when left unchecked, increases exponentially and hence doubles itself at regular intervals, was brought to prominence by Thomas Malthus in his famous work *Essay on the Principle of Population* (1798). Furthermore, the issue of exponential growth had received considerable discussion in France both before and after Malthus. For example, Auxiron (1766) and Isnard (1781) both suggested that under favorable conditions a human population could increase by 1/30 each year (see Spengler, 1965, for a commentary). This accords well with the result of Malthus that such a population would double in 25 years (1798, p. 21). In addition, Laplace had discussed the issue and in the Introduction to the second edition of his *Théorie Analytique des Probabilités* writes:

> It is difficult to evaluate the *maximum* growth of population: it seems, according to some observations, that under favorable circumstances, the population of the human species might double every 15 years. It is estimated that in North America the period of this doubling is 25 years. Under these circumstances, population, births, marriages, mortality, in short everything, increases in the same geometric progression, the constant ratio of the consecutive terms of which is obtained by observation of the annual births at two periods.

A modern analysis of the branching process model reveals that the ratio of successive generations stabilizes in the sense that, conditional on nonextinction, it converges in probability to the mean of the offspring distribution provided that this mean is greater than or equal to unity (Nagaev, 1967). Such a discussion, however, would have been well beyond possibility in Bienaymé's time.

5.10. The Bienaymé–Chebyshev Inequality

The Bienaymé–Chebyshev Inequality arises in the paper of Bienaymé (1853c) in support of the population variance as an optimal measure of dispersion.[9] He supports a *verbal* argument, which deduces from the fact that the variance σ^2/N of a sample mean $\bar{X}$ of N independently and identically distributed random variables approaches zero as $N \to \infty$, that $\bar{X}$ more closely approximates the common population mean μ, by a *mathematical* one which produces the celebrated inequality in the form: for any $t > 0$,

$$(5.8) \qquad Pr((\bar{X} - \mu)^2 \geq t^2\sigma^2) \leq 1/(t^2N).$$

In the sense mentioned, therefore, Bienaymé also has a claim to the discovery of the general Weak Law of Large Numbers, at least for identically distributed random variables; our earlier assessment of this (Heyde and Seneta, 1972, §3) was incomplete.

We repeat here, with a little expansion, our earlier historical résumé concerning the inequality. If $E(X^2) < \infty$ and $EX = \mu$, then the inequality:

$$(5.9) \qquad Pr(|X - \mu| \geq \epsilon) \leq (\text{Var } X)/\epsilon^2, \quad \text{any } \epsilon > 0,$$

is commonly understood in probability theory as the Chebyshev Inequality,[10] and less commonly as the Bienaymé–Chebyshev Inequality. If X_i, $i = 1, \ldots, N$, are independent random variables with $E(X_i^2) < \infty$ for each i, and $S_N = X_1 + X_2 + \ldots + X_N$, then since the Bienaymé Equality[11] asserts that:

$$\text{Var } S_N = \sum_{i=1}^{N} \text{Var } X_i,$$

it follows from (5.9) that:

$$(5.10) \qquad Pr(|S_N - ES_N| \geq \epsilon) \leq \sum_{i=1}^{N} (\text{Var } X_i)/\epsilon^2.$$

If in addition all the X_i are identically distributed, (5.10) becomes:

$$(5.11) \qquad Pr(|S_N - ES_N| \geq \epsilon) \leq N\sigma^2/\epsilon^2,$$

where $\sigma^2 = \text{Var } X_i$. If we put $S_N/N = \bar{X}$ and $\mu = EX_i = E\bar{X}$, and $\epsilon = Nt\sigma$, then (5.11) becomes (5.8). Clearly, when $N = 1$, (5.8) reverts to (5.9), so (5.8), (5.9), (5.10), and (5.11) are all equivalent.

---

[9] See §4.6.

[10] There is another, analytical, inequality associated with the name of Chebyshev (Mitrinović and Vasić, 1974). In a probabilistic context it gives sufficient conditions for positive (or negative) correlation between two random variables.

[11] See §4.6.

The (inhomogeneous) version (5.10) was obtained by Chebyshev (1867) and published simultaneously in French and Russian.[12] The present article of Bienaymé (1853c) was reprinted in Liouville's journal, immediately preceding the French version; there is a small editorial comment by Liouville on the reprinted paper, but this does not contain any information concerning the inequality. Only the first (27 May, 1874) of the three letters by Bienaymé to Chebyshev[13] has any bearing on the inequality, in referring to Chebyshev's (1874) paper, in which Chebyshev states:

> The simple and rigorous demonstration of Bernoulli's law to be found in my note entitled: *Des valeurs moyennes,* is only one of the results easily deduced from the method of M. Bienaymé, which led him, himself, to demonstrate a theorem on probabilities, from which Bernoulli's law follows immediately. . . .

The reader wishing to acquaint himself with Chebyshev's two papers directly (at least Russian versions thereof, in the second instance) should consult his *Oeuvres* or Chebyshev (1955). A good brief account of the direct interaction between Bienaymé and Chebyshev, to which topic we shall return shortly, is given by Sleshinsky (1892).

Perhaps Chebyshev's relative claim to priority is reasonably fairly expressed by Maistrov's (see §1.3) source, Chebyshev's illustrious pupil, Markov (1924, p. 92), in his famous textbook:

> We associate with this remarkable simple inequality, the two names of Bienaymé and Chebyshev on account of the fact that, while it was first clearly stated and proved by Chebyshev, the essential idea of proof was indicated at a substantially earlier time by Bienaymé, in whose memoir one can find the inequality itself, within the framework of some particular assumptions.

We shall, however, see below that Markov's claim as to restrictions in Bienaymé's proof of the inequality has no substance. On the other hand, there is no dispute that Chebyshev's proof is far from general. Markov's sentiments have since been reiterated by Soviet probabilists as eminent as Bernstein (1945), who adds that the circumstance somewhat detracts from the priority of Chebyshev, and that:

> Having acquainted himself with the memoir of Bienaymé [in 1867], Chebyshev, evidently, found ideas close to his own.

Markov, in fact, was ever a champion of Bienaymé's claims; in 1914 he wrote a short letter to an editor, which reads as follows:

> Dear Mr. Editor! In the note of Mr. I. Sh. entitled "In memory of Jacob Bernoulli" (*Strakhovoe Obozrenie,* 1914, No. 1, pp. 16–17) there is an incom-

[12]See §1.4.
[13]See §1.4.

plete, and therefore incorrect, reference to the memoir of Bienaymé: "Considérations à l'appui de la découverte de Laplace . . . ," which I humbly request to
correct, since, evidently, it derives from my paper.

Bienaymé's memoir should not be referred to as of the year 1867, when it was
reprinted in Liouville's journal preceding Chebyshev's memoir; but as of the
year 1853, when it was first printed in *Comptes Rendus,* vol. **37,** as I indeed
remarked in my paper.

My paper may be found in the *Vestnik Opitnoi Fiziki i Elementarnoi Matema-
tiki* for the year 1914 (No. 603).

There are two further aspects to be mentioned as regards Bienaymé's
contribution in respect of the inequality (5.8). First, and this does not
appear to have been previously noticed, his proof—in contrast to Chebyshev's—is short, simple, and indeed, that frequently used in modern
courses for direct demonstration of the inequality. Second, the proof is
associated with the "Method of Moments." In order to expand on these
aspects, we first give Bienaymé's proof, with some notational and verbal
contraction.

Suppose $\bar{X} - \mu$ is to be contained within the limits $\pm \sqrt{t^2\sigma^2}$. After N
observations, the variance of $\bar{X}$ is σ^2/N. The sample points of $(\bar{X} - \mu)^2$
outside the limit $t^2\sigma^2$, contribute only a fraction of the variance σ^2/N, i.e.,
the contribution is $f\sigma^2/N$. On the other hand, if p is the probability
corresponding to these sample points, then clearly this fraction of the
variance exceeds $pt^2\sigma^2$, since $t^2\sigma^2$ is then the minimal value of $(\bar{X} - \mu)^2$, so
that:

$$f\frac{\sigma^2}{N} = \frac{pt^2\sigma^2}{\theta},$$

where θ, like f, is a fraction between zero and one. Thus,

(5.12) $p = \theta f/\{t^2N\}$,

[which is, due to printer's error, inverted in Bienaymé's paper the first—
but not the second—time it appears, but only in the 1867 reprinting]. Then
(5.12) is equivalent to (5.8), since $0 \leq \theta f \leq 1$.

Chebyshev must have spent quite a deal of time searching for the
essence of this proof. Finally, we have his following formulation (Chebyshev, 1874) of the "method of Bienaymé":

. . . —the illustrious savant gives a method which merits very specific attention.
This method consists in the determination of the value of the integral $\int_0^a f(x)\, dx$

by the values of the integrals $\int_0^A f(x)\, dx,\ \int_0^A xf(x)\, dx,\ \int_0^A xf(x)\, dx$ . . . where A
$> a$ and $f(x)$ is an unknown function supposed only to remain positive between
the limits of integration.

If we take $f(x)$ to be the density of a nonnegative random variable Y, and A the upper limit ($\leq \infty$) of its sample space, the generalized Bienaymé–Chebyshev Inequality is:

$$(5.13) \quad \int_0^a f(x)\, dx = Pr(Y \leq a) \geq 1 - \frac{E(Y^r)}{a^r} = 1 - a^{-r} \int_0^A x^r f(x)\, dx,$$

which illustrates Chebyshev's statement, and from which the ordinary inequality (5.9) follows by putting $Y = |X - \mu|$, $r = 2$.

Chebyshev later used this method to prove the first version of the Central Limit Theorem for sums of independently but not identically distributed summands (Chebyshev, 1887); and it was quickly taken up and generalized by Markov (1907; see also 1912, p. 218) (though Chebyshev's other student, Liapounov, 1900, 1901, used the alternative method of characteristic functions).

The method later became known as the Method of Moments, so Bienaymé has some claim to be its founder, though Soviet authors tend to ascribe it to Chebyshev. This has some justification; Chebyshev's thinking on the use of the method, of which our above assessment is somewhat facile, is best illustrated by "expansion"-type arguments, as in his 1887 paper.

The inequality (5.13) with $r = 1$ is now often known as Markov's Inequality[14] and is the means now often alternatively used to prove (5.9), by putting $Y = (X - \mu)^2$. This, then, illustrates how Bienaymé's proof has diffused down to our times, and to first courses in probability and statistics.

5.11. A test for randomness

A problem which often occurs in sampling is that of randomness. For example, doubts may arise owing either to the possibility of defects in the sampling or to changes taking place while the sampling is proceeding. Alternatively, it may be of importance to test the residuals of a time series after removing a trend. Bienaymé (1874, 1875) proposed a simple test for such situations which involves the counting of the number of local maxima and minima, i.e., turning points, in the series. This is a nonparametric test and represents one of the earliest uses of a nonparametric test of a nontrivial character. The important subject bibliography, Savage (1962), lists only one substantive earlier application, that of Arbuthnot (1710), who had applied a test rather like the sign test in his paper *An argument for*

---

[14]The origin of this terminology is obscure; it may date to Markov (1884), whence the inequality is immediately deducible. This paper, a sequel to Chebyshev (1874), is, likewise, not probabilistic in character. However, Romanovsky (1925) notes that the inequality occurs in probabilistic form in the third edition of Markov's textbook.

divine providence, taken from the constant regularity observed in the births of both sexes. Of course calculations of probabilities of runs had been made by various early workers in probability such as Moivre (1718), Simpson (1740), Condorcet (1785), and Laplace (1812), but the first to develop a nonparametric test based on runs appears to have been Quetelet (1852, 1854) where tendency towards rain from one day to the next was tested against an alternative which we would now call a two-state Markov chain. This work has been discussed by Stigler (1975b).

The setup for Bienaymé's work is as follows. A turning point is a value that is either greater or less than its two neighboring values and the interval between two turning points is called a phase. In his paper of 1874, Bienaymé states that the number of phases, complete and incomplete, in a sequence of N observations from a continuous distribution is approximately Normally distributed about a mean of $(2N - 1)/3$ with variance $(16N - 29)/90$. His paper actually gives a slightly different form as a consequence of using a nonstandard form of Normal approximation; we have made the appropriate transformation into modern form. Bienaymé gives no explanation of his reasoning, only a remark about the avoidance of repeated values and a comment about the mean in the case when only two values are possible, which is $(N + 1)/2$. The paper of 1875 gives no further insight into the theory but provides six interesting applications of his result. The first three of these are on astronomical data, and the last three on daily results of loan and bond sales as reported in the newspapers. The first astronomical application concerns data of Chapelas in *Comptes Rendus* **81** (1875) on right ascensions of the commencement trajectories of the shooting stars of 10 April; the second concerns results in a note of Le Verrier, also in *Comptes Rendus* **81** (1875) on differences between observations made at Greenwich and in Paris on the heliocentric longitude of Saturn; and the third concerns a time series of observations on the right ascension and declination of Olbers's comet as reported by Bessel (1812–1813). These were, however, by no means the first tests of randomness in an astronomical context; such problems date back at least to Michell (1767) (see Todhunter, 1865, §§617–622).

Bienaymé's results aroused immediate interest and the paper of Bertrand (1875) begins as follows:

> M. Bertrand, in view of the elegant theorem enunciated by M. Bienaymé, proposes the following elementary proof of it. . . .

Bertrand, however, only concerns himself with the result for the mean which is simple by comparison with the results for the variance and the asymptotic Normality. Nevertheless, he utilizes the method of indicator variables which provides a convenient modern approach to the problem in general. There is of course no clue to the reasoning which Bienaymé had used.

We note that three consecutive observations are needed to form a turning point. Then, following a modern style of indicator approach, we consider the set of observations u_1, u_2, . . . , u_N and define indicator variables X_i, $1 \le i \le N - 2$, by:

$$X_i = \begin{cases} 1 & \text{if } u_i < u_{i+1} > u_{i+2} \text{ or } u_i > u_{i+1} < u_{i+2}, \\ 0 & \text{if otherwise.} \end{cases}$$

The number of turning points is then simply:

$$T = \sum_{i=1}^{N-2} X_i,$$

and the number of phases, complete and incomplete, is one more than the number of turning points. The X_i are of course identically distributed in the case of a random series and $\{X_i\}$ forms a portion of a stationary process which is 2-dependent (i.e., variables X_i and X_j whose indices differ by more than 2 are independent). Furthermore, in the case of a random series there will be a turning point in 4 out of the 6 permutations of 3 observations which leads to:

$$EX_i = P(X_i = 1) = \frac{2}{3},$$

and

$$ET = \frac{2}{3}(N - 2).$$

This is the extent of Bertrand's reasoning.

The computation of the variance and higher moments of T is a combinatorial exercise of increasing complexity. In order to find ET^2 it is necessary to calculate $E(X_iX_{i+1})$ $(= 5/12)$, $E(X_iX_{i+2})$ $(= 9/20)$, and $E(X_iX_{i+k})$ for $k > 2$ $(= 4/9$ easily since X_i and X_{i+k} are independent for $k > 2)$, whence we find:

$$\text{Var } T = E(T - ET)^2 = (16N - 29)/90.$$

It was principally the combinatorial aspects of Bienaymé's result that attracted attention with due acknowledgement: thus Netto (1901, pp. 106–116) summarizes the paper of Bienaymé (1874) and a long series by André published between 1879 and 1896 which includes a recursion formula for the number of phases, including the incomplete phases before the first and after the last turning points, while mention is also made of Bienaymé in the more recent paper of Schrutka (1941).

On the statistical side, while the full force of Bienaymé's result as a probability limit theorem is available in the statistical literature, its origins seem to have been largely forgotten. This is despite an early report by

Mansion (1881) and a citation in the bibliographical work of Merriman (1877) in which the following inaccurate comment is included:

> If a series of observations be arranged in order of the measurements, there are certain maxima and minima whose probable number and position are given by the theorem.

Bilham (1926) treats the mean number of turning points, but without reference to Bienaymé. Reference to Bienaymé (1874, 1875) does appear in an historical appendix concluding the paper of Wallis and Moore (1941b), who deal in part in their preceding theory with precisely the problem considered by Bienaymé; this paper is a much-expanded version of Wallis and Moore (1941a), which does not refer to Bienaymé. This material is given later by Kendall (1955, pp. 124–125) and Kendall and Stuart (1968, pp. 351–352) but without attribution; and a test of randomness based on Bienaymé's result is called a turning-point test. Asymptotic Normality is inferred from an investigation of skewness and kurtosis, as in Wallis and Moore (1941b). Indeed, after tedious calculation,

$$E(T - ET)^3 = -16(N + 1)/945,$$
$$E(T - ET)^4 = (448N^2 - 1976N + 2301)/4725,$$

so that the skewness:

$$\gamma_1 = E(T - ET)^3/[E(T - ET)^2]^{3/2} = 0 \ (N^{-1/2}),$$

and the kurtosis:

$$\gamma_2 = \{E(T - ET)^4/[E(T - ET)^2]^2\} - 3 = 0(N^{-1}),$$

as $N \to \infty$. These results give an indication of likely asymptotic Normality, but certainly not a proof. A modern proof could use the Central Limit Theorem for stationary m-dependent processes. Alternatively, a finite stationary irreducible Markov chain can be constructed and the asymptotic behavior obtained from the Central Limit Theorem for random variables defined on such a chain. All these methods would have been well beyond the scope of Bienaymé in 1874. Even the use of the skewness and kurtosis to measure departure from Normality dates from the work of Pearson. The concept of kurtosis was introduced in Pearson (1906); see Walker (1931) for some perspective. It would indeed be interesting to know something of Bienaymé's reasoning. The work represents a remarkable achievement for its time and it should additionally be noted that Bienaymé turned 78 years of age in April of 1874.

Kendall and Stuart (1968) remark that the turning-point test is reasonable as a test against cyclicity but poor as a test against trend. This is to be expected because turning is a local property and is not much influenced by a modest trend. Nevertheless, this points to an explanation for the essential disappearance of Bienaymé's name from the literature. Beginning with André (1884), there commenced an investigation of *phase lengths,* as distinct from Bienaymé's *number of phases.* This has evolved into a large body of literature concerned with runs and leading to tests which have wider use than that of Bienaymé. Notable early contributions on the subject of phase lengths were Kermack and McKendrick (1936, 1937) and Wallis and Moore (1941b).

In a random series the mean phase length is approximately 1.5 with a variance of approximately 0.56; further investigation is suggested in a series of more than 10 terms whenever the mean phase length is 2 or more. Wolfowitz (1944) demonstrated asymptotic Normality of the number of phases and Levene (1952) studied the power of tests based on phases. Gleissberg (1945) tabulated the distribution of the number of phases for sample sizes $N \leq 25$. Other contributors at about the same period were Moore and Wallis (1943), Levene and Wolfowitz (1944), and M. G. Kendall (1945).

6. Miscellaneous writings

If the law supposes that, . . . the law is
an ass. . . .

C. Dickens

6.1. A perpetual calendar

Bienaymé's breadth of interests has been pointed out in §1.3. In this section we note one of his more unusual pieces of work, namely his presentation (1841) to the Academy of Sciences of a perpetual calendar. Unfortunately no details are given save that the calendar was of the chart kind, involving marking on the same plan the date and day of the week.

Various chart calendars had been published in the late eighteenth century, for example the *Secular Diary* of D. Barlow which covered all dates from 1601 to 1900, inclusive. Bienaymé's calendar presumably offered longer periods than its competitors or provided a construction of greater simplicity.

More elaborate perpetual calendars of a mechanical kind were produced in the nineteenth century. For example, *A Perpetual Almanack and Calendar for the Investigation of Dates* published by John Gardiner in Edinburgh in 1868 was printed on stiff paper and had a type of sliderule calculator as an integral part. It could be used to compute all past and future dates and days.

6.2. The alignment of houses

In 1851, Bienaymé published a satirical dialog extending to 24 pages entitled *De la mise à l'alignement des maisons*. This appeared as a publication of *La Revue Municipale* and concerns iniquities of the land resumption system of the time in Paris. The lively dialog is of a *Catch 22* style and would make an amusing play. It is between an individual, whose property had been subjected to a resumption order, and an official of the administration. The essence of the trouble is clearly stated in the following quotation (Question, p. 22):

> When the French authorities consider that they should have a street widened, there are, in their view, two methods open to them.
>
> If the widening is a matter of urgency, they purchase the houses, the demolition of which is necessary. This purchase is made by *compulsory expropriation in the public interest*. A jury is set up which establishes the value of each house, and the price thus fixed is paid to the various owners before the authorities take possession of the houses.
>
> If, on the other hand, the widening is not urgent, the authorities are content to draw up a plan of the street and the houses on it. On this plan a line is drawn indicating the portion of each house necessary for the widening of the street.
>
> From this moment, repairs cannot be made to the portions of each of these houses which lie outside the line.
>
> Due to lack of repairs, the result is that the houses reach a decrepit state more rapidly than if they had been repaired.
>
> When one of these houses threatens to collapse, the owner is obliged to demolish it and to rebuild it only on the new line.
>
> He is paid only the value of the land. Thus, in the first case the owner receives not only the price of the land which is taken from him, but also of the building which occupied this land.
>
> In the second case he receives only the price of the land.

The article is carefully documented and complete with appropriate legal references. It is quite clear that Bienaymé had strong feelings on the subject; so strong that it is not unreasonable to surmise that his house was involved. Indeed, one of his biographers, Sagnet, refers to his abhorrence of the law.[1] It should be noted that Bienaymé was not formally employed at the time; he had left the civil service in 1848 and was elected to the Academy of Sciences in 1852.

The program of street widening and associated public works was the responsibility of the Prefect of the Seine Department. At the time of Bienaymé's article the Prefect was Berger, who was a stopgap appointment, holding the post between 1848 and 1853. The previous Prefect was the Marquis de Rambuteau, who held the position from 1833 to 1848, and the successor to Berger was the famous Baron Haussmann, who was the incumbent from 1853 to 1870.

Rambuteau grossly underestimated the scale of developments necessary under his administration. Neither he nor Louis Philippe was aware of the effects of their moderation. Indeed, Rambuteau referred to himself as the "Surveyor Prefect" and prided himself on the 112 roads (of unspecified

---

[1]See §1.3.

length) that were opened during his administration. However, the widest of these, which bears his name, is only 40 ft in width. It is presumably the effects of Rambuteau's moderation which Bienaymé had experienced, for it is certainly not the urgent resumptions that are at issue.

The perspective of activity was to change completely during Haussmann's administration. He was responsible for the transformation of Paris from its ancient character to the one which it still largely preserves. An enormous program of public works was embarked upon. Large straight avenues were cut through the chaotic mass of small streets of which Paris was then composed. New systems of water supply and drainage were provided. Most of the private buildings on the Île de la Cité were demolished and it was given its official character. The Opéra and Les Halles markets were built. Haussmann operated with little parliamentary or other control and was able to raise enormous long-term loans to finance his activities. It was, however, increasing criticism of his handling of public money which finally led to his dismissal.

With the change from Rambuteau's to Haussmann's administration the problems of the system appear to have increased despite modifications aimed at the appearance of honesty. For background on the period see J. M. and B. Chapman (1957). The following quotation, from pp. 91–92 of that source, well evidences the abuses that were rife in the system that operated at least from 1854:

> Speculators tried to forsee the paths of new roads so that they could buy up land in advance of the administration. When it had been decided to expropriate a piece of land, the architect surveyors of the planning section . . . advised the municipal council on the compensation to be offered to the proprieter or tenant, and also fixed the price of any plots of land which could be resold and of the materials derived from the demolition. . . .
>
> There was another body even more open to corruption. This was the estate office (the bureau du domain de la ville), an administrative division outside the planning section which was authorized to arrange unofficial settlements for eviction and expropriation claims. This system had been introduced to soften opposition, when necessary at a price, and was by definition open to flagrant abuse. To protect property owners from official abuse the Supreme Court ruled that every expropriated person had the right to appeal to an expropriation jury to assess the value of his property. But this ruling gave rise to equally great abuse on the other side. Sometimes it was found that expropriated persons or the friends of interested parties sat on the expropriation jury, all naturally ready to benefit each other at the expense of the administration.

It is interesting to note that Émile Zola gives a pungent account of the abuses of land expropriation in his novel La Curée, first published in 1872 (particularly in Chapter 2). Zola undoubtedly had a strong social conscience, as evidenced by his strenuous intervention in the Dreyfus affair.

Finally, we have as an example of Bienaymé's noted facility with languages (see §1.3) some humorous Latinized verse with which he concludes the main text:

... je vous proposerais de terminer les brevets [of the administration] de la manière suivante:

Atque ego
In virtute allianciae
Chartae et voiriae
Cum isto boneto
Doctissimo,
Sed praesertim liberalissimo
Et constitutionalissimo,
Tibi dono et concedo
Facultatem et puissanciam
Muros et domos
Perçandi
Taillandi
Coupandi
Et *tout doucement* eruendi
Impunè
Et *sans payer,*
Per totam Franciae terram
—Monsieur, vous oubliez le réalignement
—Monsieur, je n'ai pas fini
Et post reedificationem
Rursùs
Et retrò
Et iterum
Et *toujours*
Retaillandi
Et recoupandi
Êtes-vous content, monsieur?
—Monsieur, permettez que je vous embrasse.

6.3. The Montyon Prize reports

The Montyon Prize for Statistics was at least until recently a prestigious annual award of the Academy of Sciences of Paris, being one of a collection of prizes for various subject areas. The Academy annually appointed a committee, usually 5 in number in Bienaymé's time, to recommend on the award of the Montyon Prize for that year. Bienaymé was appointed to the committee for the Prize of 1853 soon after his election to the Academy and was reappointed annually until the year of his death. For much of the time he was the spokesman of the committee and delivered its report to the Academy.

The reports of the committee provide a detailed critical commentary on the works which were short-listed in the competition for the prize of a particular year. They are published in full in *Comptes Rendus* and often run to 5 or more pages. A listing of the reports of the prize committees of which Bienaymé was a member is provided within Table 2.

In this section we shall discuss only the reports that relate especially to Bienaymé's own interests. These are the reports of the competitions for the years 1853, 1856, 1870, 1874, and 1876. In each case Bienaymé was the spokesman. The other reports make interesting reading on subjects of great diversity, ranging over such memoirs as *Du Dromadaire, comme bête de somme et comme animal de guerre* and *Sur le prix de revient des transports par chemins de fer,* but are outside the scope of the present account. All are notable for their erudition and perception.

The report of the competition for the Prize for the year 1853 appeared in *Comptes Rendus* **38** (1854), 133–146. Ten works were submitted in this case and five were short-listed. The prize itself was not awarded but two works were accorded a monetary award. First among these was a manuscript entitled *De l'organization des Sociétés de prévoyances ou des secours mutuels, et des bases scientifiques sur lesquelles elles doivent être établies, avec une Table de mortalité et une Table de maladie, dressées sur des documents spéciaux; publié sous la direction du comité pour la propagation des Sociétés de prévoyance* by Gustave Hubbart, secretary of the committee referred to in the title. A detailed report, running to approximately 7 pages, is given on this memoir.

The subject matter of Hubbart's article is very relevant to the work of Bienaymé which has been discussed in §2.5 and we note that it is temporally intermediate between his article (1839b) and a later group (1857, 1862a,b,c, 1865). However, there is comparatively little discussion of the issue of stability of mutual security societies in the report; the emphasis is much more on the tables.

The memoir was faulted on a number of grounds in the report. First, the methods employed were not new. Second, it was criticized for placing too much confidence in foreign statistical work and too little in French work and, indeed, for suggesting that the French work was backward. There was, in fact, rather little French experience in mutual security societies compared with that of neighboring countries prior to government encouragement in a law of 13 July, 1850. The report notes, rather happily, that various English tables cited by the author are in *"discorde flagrante"* with one another. In fact, between 21 and 65 years of age one lists 407 days of sickness, another 551, and a third, and most recent, 655. The table given in the memoir lists 386 but the author has offered no explanations.

It is suggested in the report that the results in the memoir provide confirmation of the mortality table of Deparcieux. Bienaymé of course had strong feelings on this subject (see §2.3) and his influence on the comments is obvious. It is shown in the report how an adjustment made by adding in

1911 observations which had been rejected, on the basis of incomplete data, largely reconciled the mean lifetime according to the table in the memoir with that given by the Deparcieux table.

The report concludes with a reference to the work of Didion on the operation of the Metz mutual security society for 20 years, which was to provide the basic stimulus for Bienaymé's papers (1857, 1862a,b,c, 1865).

The next report on which we shall comment is that for the Prize of 1856 which appeared in *Comptes Rendus* **44** (1857), 159–163. The Prize in this case was awarded to a book entitled *Consommations de Paris* by Armand Husson. The book by Husson was a natural successor to two previous works on the subject. The first of these was a large volume entitled *La Richesse territoriale de la France* by the famous chemist Lavoisier, published in 1791. The second was a memoir, also entitled *Consommations de Paris,* by Benoiston de Châteauneuf, first published in 1819 and later, in 1821, expanded to two volumes.

The report is lavish in its praise for the work of Châteauneuf which had remained without a successor for more than 30 years. Further, advanced age had prevented Châteauneuf from updating his work recently and he died in 1856. Other work of Châteauneuf has been mentioned earlier in our §2.2 and, apparently in a key role, in our §5.9. The work by Husson dealt only with the consumption of food, which had been the subject of the first volume of Châteauneuf's book (the second volume had dealt with industrial consumption). The type of consumption questions under consideration were the daily consumption of bread per head and the annual expenditure on food per head. In the case of the former the figures obtained are: Lavoisier (*circa* 1789), 459 grammes; Châteauneuf (1817), 497 grammes; and Husson, 510 grammes. In the case of annual expenditure, the figures are: Lavoisier 260 fr., Châteauneuf 385 fr., and Husson 480 fr. However, the report expresses considerable doubts on various issues, such as the population sizes on which the figures have been computed. There is no comment on the question of inflation.

Work of Châteauneuf also figures prominently in the report for the Prize of 1870 which appeared in *Comptes Rendus* **75** (1872), 1306–1314. The Prize in this case was awarded for a memoir entitled *l'Institut de France, ses diverses organisations, ses Membres, ses Associés et ses Correspondants,* by A. Potiquet, which provides a detailed demographic analysis of the Institute based on details of all Members and Corresponding Members elected from the time of its creation, in 1795, up till 19 November, 1869. This report includes detailed comparison of the results of Potiquet with those of Châteauneuf (1841) which had provided a similar kind of analysis using data based on members of three of the four prerevolutionary Academies and those of the Institute elected before 1 January, 1840. The old Academies were *l'Académie Française* (founded 1635), *l'Académie des Inscriptions et Belles-Lettres* (founded 1663), *l'Académie des Sciences* (founded 1666), and *l'Académie des Beaux-Arts;* they were suppressed in 1793 and became part of *l'Institut National de France* in 1795. Châteauneuf's data was, however, rather incomplete with regard to some of the less well-known early members of the Academies. He also avoided inclusion of results for the *Académie des Beaux-Arts* since the average age at entry was substantially below that of the other Academies.

The results of Châteauneuf gave mean age at election as 44 years and 2 months, with mean age at death as 68 years and 10 months. On the other hand, the later data used by Potiquet gave mean age at election as 51 years and 10 months, with mean age at death being 71 years and 5 months. Detailed demographic analyses on mortality are given in the report comparing the two collections of Academy data with the Deparcieux tables. Not surprisingly, the Academicians tended to live substantially longer than predicted by the Deparcieux tables. Châteauneuf had previously commented on this phenomenon and had been of the opinion that the tranquil life of a scholar would promote longevity. Of course it is not difficult to provide other explanations.

A system of elections in exclusive and prestigeous learned societies is often geared to an expected balance of deaths against new members rather than one of direct replacement of a deceased member as in *l'Institut de France.* Of course a fixed number of members elected per annum, arranged to achieve an essentially constant membership, requires extra careful monitoring of the statistics. For early analyses of the vital statistics of the Royal Society of London and of the National Academy of Sciences of the U.S.A., see Strachey (1872) and Pearl (1925), respectively.

In addition to the memoir by Potiquet, two other works were accorded an honorable mention in the competition for the Prize of 1870. Second among these was a brochure entitled *De l'influence de la température sur la mortalité de la ville de Montpellier,* by A. Castan, which is closely related in subject matter to Bienaymé's first paper (1829); see §2.2. The mortality data for the period 1859 to 1868 in Montpellier indicated highest mortality rates in summer for infants under 2 years of age and in winter for persons over 60 years of age.

The next report on which we comment is that for the Prize of 1874 which appeared in *Comptes Rendus* **80** (1875), 1474–1479. Here the Prize was awarded to a M. de Kertanguy for a work entitled *Sur la Mortalité parmi les assurés de la Compagnie générale* which appeared in *Journal des Actuaires français* in three parts in 1874 and 1875. The data considered in this work involved 28,000 assurance policies covering 24,699 persons of which only 3,899 were females. The report is particularly concerned with the life table constructed in de Kertanguy's memoir in comparison with that of Deparcieux and the recently published 20 English Companies table (see §2.3). In fact, the three tables are quite similar, as indicated by the following comparison of mean lifetimes remaining at different ages.

Age	de Kertanguy	Deparcieux	20 English Companies
35	30.75	30.88	31.15
40	26.95	27.47	27.57
45	23.18	23.88	23.98
50	19.75	20.38	20.51
55	16.33	17.24	17.14
60	12.95	14.25	13.99
65	10.31	11.25	11.17
70	8.07	8.63	8.68
75	6.55	6.50	6.56
80	5.60	4.75	4.93
85	3.19	3.34	3.58

This latest corrobaration of the Deparcieux table is noted with some satisfaction in the report and evidences Bienaymé's almost obsessive concern with this issue. It is no surprise to learn from this report that the actuarial calculations on which the government retirement fund were based made use of the Deparcieux table (see §1.3 and §2.5). Some detailed comments are also made in the report on the number of observations necessary to obtain specified precision of the age-specific death rates in a mortality table. The results are based on work of Laplace following on from that of Jacob Bernoulli. It is noted that most tables are based on insufficient data to provide good estimates in the tail.

A final report of interest is that for the Prize of 1876 which appeared in *Comptes Rendus,* **84** (1877), 817–825. In this case the Prize was not awarded but very honorable mention was accorded to Dr. Bertillon for his work entitled *Démographie de la France, mortalité selon l'âge, le sexe, l'état civil en chaque département, et pour la France entière.* Nevertheless, the report is quite critical of Dr. Bertillon's memoir on a number of grounds. He is criticized for his dogmatism and for the data themselves:

> It would perhaps be necessary to add that this report [of Bertillon] has led to the fabrication of population numbers which had not been obtained. This is no longer statistics.

A major point of interest in the report is a paragraph indicating a profound knowledge of classical Greek on the part of one of the committee (presumably Bienaymé, who was the spokesman; see §§1.3, 5.8). Bertillon had used the term *mortality* to mean the ratio of deaths to total population and the report continues as follows:

> M. Bertillon, having thus changed the term mortality from the meaning in which it has always been generally and naturally accepted, has felt obliged to replace it by another. He has chosen *dîme mortuaire,* which does not seem to be a fortunate choice. The invention of new words is an extremely delicate matter, and the changing of accepted senses is even worse. It would be possible to reproach him for having entitled his curious tables *demography.* The Greek word δῆμοσ is applied to the people regarded as a body politic; the people, as

composed of men and to a greater extent of women and children, is more properly designated as λαόσ. From this is derived the word λαογραφία which exists in the dictionaries, and which meant a census. The worthy Homer, expressing what La Fontaine has so well translated as *at all times the young have suffered from the foolishness of adults,* says: ὀλέχοντο δε λαοί and not δῆμοι. His line, however, could just as well have ended with the second word as with the first. On reading *Démographie de la France* without the explanation which the clear sense of M. Bertillon has brought to it, one would not think of statistics of human life, but of some political developments of France in its assemblies.

Table 2. References to Bienaymé extracted from name indexes of Comptes Rendus Hebd. des Séances de l'Académie des Sciences[1]

A number of announcements, notes, and reports not otherwise mentioned in the text may be traced from this list. The year of publication of a given volume may be located approximately by bearing in mind that **1** appeared in 1835, and two volumes per year seem to have been published thereafter: thus **34–35** (1852), **58–59** (1864), **60–61** (1865).

Sur les erreurs présumées des documents à l'aide desquels on a calculé, en France, les tables de population. I, 364.

Note sur la durée de la vie de l'homme en France, depuis le commencement du XIXe siècle. I, 417.

Notice sur un nouveau quantième perpétuel. XIII, 1103.

Mémoire sur la probabilité des erreurs d'après la méthode des moindres carrés. XXXIII, 458.

Rapport sur ce Mémoire; Rapporteur M. *Liouville*. XXXIV, 90.

M. *Bienaymé* est présenté comme un des candidats pour la place d'Académicien libre vacante par suite du décès de M. *Maurice*. XXXIV, 192.

Et pour celle d'Académicien libre vacante par suite de la morte de M. le Maréchal *Marmont*. XXXIV, 991.

M. *Bienaymé* est nommé à la place d'Académicien libre en remplacement de M. le Maréchal *Marmont*. XXXV, 10.

Décret approuvant cette nomination. XXXV, 37.

M. *Bienaymé* est nommé Membre de la Commission du prix de Statistique. XXXVI, 850; XXXVIII, 807; XL, 102; XLII, 991; XLIV, 318; XLVI, 796; XLVIII, 620; L, 845; LII, 1062; LIV, 361; LVI, 126; LVIII, 42; LX, 518.

Sur les différences qui distinguent l'interpolation de M. *Cauchy* de la méthode des moindres carrés et qui assurent la supériorité de cette méthode. XXXVCI, 5.

Remarques à la suite d'une communication de M. *Cauchy* à l'occasion de ce Mémoire, XXXVII, 68.

Remarques sur l'ordre dans lequel ont paru deux parties d'un Mémoire de M. *Cauchy*. XXXVII, 197.

---

[1]*Table Générale: Tome* **1–31**, *Tome* **32–61**, *Tome* **62–91**.

Table 2 139

Remarques relatives à un nouveau Mémoire de M. *Cauchy* sur les mêmes questions. XXXVII, 206.

Considérations à l'appui de la découverte de *Laplace* sur la loi de probabilité dans la méthode des moindres carrés. XXXVII, 309.

Rapports sur le concours pour le prix de Statistique. XXXVIII, 133; XL, 38; XLIV, 159; XLVI, 267; XLVIII, 489; L, 202; LIII, 1131; LV, 937.

Remarques à l'occasion d'un Mémoire de M. *Didion* où se trouve le calcul du taux des pensions de la Société de secours mutuels de Metz pour la période de 1854 à 1859. XLIV, 573.

Rapport sur une "Table de divisions," de M. *Ramon Picarte*. XLVIII, 328.

Lettre à M. *Chasles* sur l'aurore boréale du 1er octobre 1859. XLIX, 481.

M. *Bienaymé* présente, au nom de M. *Willich,* un exemplaire des "Tables populaires." L, 188.

M. *Bienaymé* présente un ouvrage de M. *Picarte* intitulé: "La division réduite à une addition." LI, 889.

Remarques sur les Sociétés de secours mutuels. LIV, 536.

Remarques à l'occasion d'une réclamation de la *Société de prévoyance et de secours mutuels de Metz,* relative à l'inexactitude de quelques-unes des données sur lesquelles M. *Bienaymé* s'est appuyé. LIV, 796.

M. *Bienaymé* maintient l'exactitude des données dont il s'agit et que lui a fournies M. *Didion*. LIV, 889.

M. *Bienaymé* est nommé Membre de la Commission chargée de préparer une liste de candidats pour la place d'Académicien libre vacante par suite du décès de M. *Du Petit-Thouars*. LX, 325.

M. *Bienaymé* présente, au nom de M. *Didion,* une brochure sur le calcul des pensions dans les Sociétés de prévoyance. LX, 415.

Rapport sur le Concours de *Statistique* de l'année 1865. (Travaux de MM. *Chenu, Poulet, Sistach, Saintpierre*.) LXII, 478.

Rapport sur le Concours de *Statistique* de l'année 1866. (Travaux de MM. *Brochard, Parchappe, Le Fort, Plessier, Girard de Cailleux*.) LXIV, 443.

M. *Bienaymé* présente à l'Académie le "Compte rendu statistique de l'Administration des hôpitaux de Rome pour 1865," et indique le caractère de cet ouvrage. LXV, 1053.

Rapport sur le Concours du prix de *Statistique* de l'année 1867. (Travaux de MM. *Eugène Marchand, Vacher, Bergeron, Blanchet, Beauvisage*.) LXVI, 925.

Rapport sur le concours du prix de *Statistique* de l'année 1868. (Travaux de MM. *Bérigny, Ébrard, Fayet, Charpillon, Rambosson*.) LXVIII, 1349.

Rapport sur le Concours du prix de *Statistique* de l'année 1869. (Travaux de MM. *Chenu, Magué* et *Poly, Bontemps*.) LXXI, 94.

Traduction de deux passages de *Stobée* inexpliqués jusqu'ici. LXXI, 460.

Rectification des listes d'articles détachés de M. *Cauchy,* publiées dans deux catalogues différents, et restitution à M. *Cournot* de quelques-uns de ces articles. LXXII, 25.

Rapport sur le prix de *Statistique* de l'année 1870. (Travaux de MM. *A. Potiquet, A. Thévenot, A. Castan.*) LXXV, 1306.

Rapport sur le prix de *Statistique* de l'année 1871. (Travaux de MM. *E. Cadet, Ély.*) LXXV, 1349.

Rapport sur le Concours du prix de *Statistique* pour 1872. *(Revue maritime et coloniale.)* LXXIX, 1543.

Rapport sur le Concours du prix de *Statistique* pour 1873. (Travaux de MM. *F. Lucas, Sueur, H. Bertrand.*) LXXIX, 1611.

Rapport sur le Concours du prix de *Statistique* de la fondation Montyon pour 1874. (Travaux de MM. *de Kertanguy, de Saint-Génis, Loua.*) LXXX, 1474.

Application d'un théorème nouveau du Calcul des probabilités. LXXXI, 417.

Rapport sur le Concours du prix de *Statistique* pour 1875. (Travaux de MM. *Chenu, Borius, Maher, Ricoux, Lecadre, Trémeau de Rochebrune.*) LXXXI, 1321.

Rapport sur le Concours du prix de *Statistique* pour 1876. (Travaux de MM. *Bertillon, Heuzé, G. Delaunay.*) LXXXIV, 817.

Rapport sur le Concours du prix de *Statistique* pour 1877. (Travaux de MM. *E. Yvernès, T. Loua, Distère, Puech.*) LXXXV, 167.

M. *Bienaymé* est nommé membre de la Commission chargée d'une révision de la Table de mortalité de *Deparcieux*, LXXXVI, 226.

M. *Bienaymé* est nommé membre de la Commission des prix de *Statistique* de la fondation Montyon: pour l'année 1866, LXII, 741; pour l'année 1867, LXIV, 1271; pour l'année 1868, LXVI, 792; pour l'année 1869, LXIX, 42; pour l'année 1870, LXXI, 215; pour l'année 1872, LXXV, 1606; pour l'année 1874, LXXIX, 200; pour l'année 1875, LXXX, 997; pour l'année 1876, LXXXII, 1141; pour l'année 1877, LXXXIV, 1067; pour l'année 1878, LXXXVI, 751.

M. *Bienaymé* est nommé membre de la Commission chargée de présenter une liste de candidats pour la place d'Académicien libre, laissée vacante par le décès de M. le maréchal *Vaillant*. LXXXVI, 409.

Et de la Commission chargée de présenter une liste de candidats pour la place d'Académicien libre, devenue vacante par la démission de M. le comte *Jaubert*. LXXXVI, 997.

M. *le President* annonce la mort de M. *Bienaymé*, décédé le 19 octobre 1878. LXXXVII, 569.

Note de M. *de la Gournerie* sur les travaux de M. *Bienaymé*. LXXXVII, 569.

Bienaymé's publications

Titles of articles in *Soc. Philomat. Paris Extraits* are usually supplemented in the list below by the title of the same article as published in *L'Institut,* marked (I.), or Bienaymé's own title, marked (B.), from Bienaymé (1852b).

(1829) Mortalité des nouveaux-nés. *Moniteur Universel,* 18 April, 564.

(1835) Sur les erreurs présumées des documents à l'aide desquels on a calculé, en France, les tables de population *C. R. Acad. Sci., Paris,* **1,** 364.

(1837a) De la durée de la vie en France. *Annales d'Hygiène Publique, Paris,* **18,** 177–218.

(1837b) Naissances des enfants dans le mariage. Sur le nombre proportionnel des enfants dans la première année d'une grande quantité de mariages (B.). *Soc. Philomat. Paris Extraits,* Ser. 5, 153–155. *L'Institut,* **223,** Vol. 6 (1838), 18.

(1838a) Probabilité des jugements et des témoignages. Sur les erreurs de la méthode suivie dans le calcul de la probabilité des témoignages et des jugements (B.). *Soc. Philomat. Paris Extraits,* Ser. 5, 93–96. *L'Institut,* **235,** Vol. 6, 207–208.

(1838b) Mémoire sur la probabilité des résultats moyens des observations; démonstration directe de la règle de Laplace. *Mém. pres. Acad. Roy. Sci. Inst. France,* **5,** 513–558.

(1839a) Théorème sur la probabilité des résultats moyens des observations. Sur la probabilité des résultats moyens lorsque les causes sont variables durant les observations (B.). *Soc. Philomat. Paris Extraits,* Ser. 5, 42–49. *L'Institut,* **284,** Vol. 7, 187–189.

(1839b) Effets de l'intérêt composé. *Soc. Philomat. Paris Extraits,* Ser. 5, 60–65. *L'Institut,* **286,** Vol. 7, 208–209.

(1840a) Application à la statistique médicale. Sur les difficultés de l'application du calcul des probabilités à la statistique médicale (B.). *Soc. Philomat. Paris Extraits,* Ser. 5, 10–13. *L'Institut,* **322,** Vol. 8, 78–79.

(1840b) Calcul des probabilités. Sur ce qu'on doit entendre par les mots: Constance des résultats moyens (B.). *Soc. Philomat. Paris Extraits,* Ser. 5, 18–22. *L'Institut,* **326,** Vol. 8, 111–112.

(1840c) Calcul des probabilités. Solution d'une problème du calcul des probabilités se rattachant avec élections (I.). Sur la probabilité de la distribution d'une majorité d'électeurs entre un grand nombre de collèges (B.). *Soc. Philomat, Paris Extraits,* Ser. 5, 23–26. *L'Institut,* **327,** Vol. 8, 119–120.

(1840d) Probabilités. Principe nouveau du calcul des probabilités avec ses applications aux sciences d'observation (I.). Sur la constance des causes, conclue des effets observés (B.). *Soc. Philomat. Paris Extraits*, Ser. 5, 37–43. *L'Institut*, **333,** Vol. 8, 167–169.

(1840e) Quelques propriétés des moyens arithmétiques de puissances de quantités positives. *Soc. Philomat. Paris Extraits*, Ser. 5, 67–68. *L'Institut*, **342,** Vol. 8, 216–217.

(1841) Notice sur un nouveau quantième perpétuel. *C. R. Acad. Sci., Paris*, **13,** 1103.

(1843) [Untitled communication in the area: Histoire des mathématiques, concerning Pascal]. *Soc. Philomat. Paris Extraits*, Ser. 5, 49–50. *L'Institut* (details not known).

(1845) De la loi de multiplication et de la durée des familles. *Soc. Philomat. Paris Extraits*, Ser. 5, 37–39. *L'Institut*, **589,** Vol. 13, 131–132. Reprinted at the conclusion of D. G. Kendall (1975).

(1851) *De la mise à l'alignement des maisons*. Publications de la Revue municipale. Dondey-Dupré, Paris (24 pp.).

(1852a) Mémoire sur la probabilité des erreurs d'après la méthode des moindres carrés. *Liouville's J. Math. Pures Appl.*, (1) **17,** 33–78. Also (1858) *Mém. pres. Acad. Sci. Inst. France* (2) **15,** 615–663.

(1852b) *Notice sur les travaux scientifiques de M.I.-J. Bienaymé*. Bachelier, Paris (4 pp.).

(1853a) Sur les différences qui distinguent l'interpolation de M. Cauchy de la méthode des moindres carrés, et qui assurent la supériorité de cette méthode. *C. R. Acad. Sci., Paris*, **37,** 5–13. *Liouville's J. Math. Pures Appl.*, (1) **18,** 299–308.

(1853b) Remarques de M. Bienaymé à l'occasion des Notes inserées par M. Cauchy dans les Comptes Rendus de deux des séances précédentes. *C. R. Acad. Sci., Paris*, **37,** 197–198.

(1853c) Considérations à l'appui de la découverte de Laplace sur la loi de probabilité dans la méthode des moindres carrés. *C. R. Acad. Sci., Paris*, **37,** 309–324. Also (1867) *Liouville's J. Math. Pures Appl.*, (2) **12,** 158–176.

(1855) Sur un principe que M. Poisson avait cru découvrir et qu'il avait appelé Loi des grands nombres. *C. R. Acad. Sci. Morales et Politiques* (3) **11,** 379–389. Also (1876) *J. Soc. Statist. Paris*, 199–204.

(1857) [Untitled remarks on a report by M. Didion concerning the calculation of pensions by the Société de Metz, 1855–1859]. *C. R. Acad. Sci., Paris*, **44,** 573.

(1858) [see Chebyshev, 1855].

(1862a) Remarques sur les Sociétés de Secours mutuels. *C. R. Acad. Sci., Paris*, **54,** 536–537.

(1862b) [Untitled remarks following the paper Société de Metz, 1862]. *C. R. Acad. Sci., Paris*, **54,** 796.

(1862c) Remarques sur les Sociétés de Secours mutuels. *C. R. Acad. Sci., Paris*, **54,** 889–891.

(1865) [Untitled remarks in connection with the presentation of a pamphlet entitled ''Calcul des pensions dans les Sociétés de prévoyance'' by M. Didion] *C. R. Acad. Sci., Paris*, **60,** 415.

(1870) Traduction de deux passages de Stobée inexpliqués jusqu'ici. *C. R. Acad. Sci., Paris,* **71,** 460–462.

(1871) Rectification de listes d'articles détachés de M. Cauchy, publiées dans deux Catalogues différents, et restitution à M. Cournot de quelques—uns de ces articles. *C. R. Acad. Sci., Paris,* **72,** 25–29.

(1873) Rapport sur le Concours pour le prix de Statistique, fondation Montyon. (*C. R. Acad. Sci., Paris,* **75,** pp. 1306, 1349—séance du 25 nov. 1872). *Liouville's J. Math. Pures Appl.,* (2) **18,** 164–179.

(1874) Sur une question de probabilités. *Bull. Soc. Math. France,* **2,** 153–154.

(1875) Application d'un théorème nouveau du calcul des probabilités. *C. R. Acad. Sci., Paris,* **81,** 417–423. *Bull. Math. Astr.,* **9,** 219–225. Apparently first published (1874) in *Bull. Soc. Math. France,* **2.**

Three letters to P. L. Chebyshev, dated respectively 27 May, 1874; 24 December, 1874; and 14 August, 1875. In Russian translation, on pp. 439–441 of Chebyshev (1951).

Bibliography

Abbé, E. (1863) Über die Gesetzmässigkeit der Verteilung der Fehler bei Beobachtungsreihen. *Gesammelte Abhandlungen*, **2**, 55–81.

Adams, W. J. (1974) *The Life and Times of the Central Limit Theorem*. Kaedmon, New York.

Adrain, R. (1808) Research concerning the probabilities of the errors which happen in making observations. *The Analyst, or Mathematical Museum* (Philadelphia), **1**, 93–109.

Aitken, A. C. (1957) *Statistical Mathematics*. (8th ed.). Oliver and Boyd, Edinburgh.

Albertsen, K. (1973) Sloegternes Uddøen—Eet Matematisk-Statistik Problem. (On the extinction of families; in Danish). *Personalhistorisk Tidsskrift*, Ser. 16, **1**, 109–130.

Ampère, A. M. (1802) *Considérations sur la Théorie Mathématique de Jeu*. Périsse Frères, Lyons and Paris.

Anderson, O. (1954) *Probleme der Statistischen Methodenlehre in den Sozialwissenschaften*. (1st & 2nd eds.; 4th ed.: 1962) Physica, Würzburg.

André, D. (1884) Étude sur les maxima, minima et séquences des permutations. *Ann. Éc. Norm. Sup.*, (3) **1**, 121–134.

Arbuthnot, J. (1710) An argument for divine providence, taken from the constant regularity observed in the births of both sexes. *Phil. Trans. Roy. Soc. London*, **27**, 186–190.

Arnauld, A., and Nicole, P. (1970) *La Logique ou l'Art de Penser*. (Reproduction of first anonymous Paris edition of 1662) Georg Olms, Hildesheim.

Auxiron, C. F. J. (1766) *Principes de tout gouvernement ou examen des causes de la splendeur ou de la faiblesse de tout état considéré en lui-même et independamment des moeurs*. J.-T. Hérissant, Paris.

Bachelier, L. (1914) *Le Jeu, La Chance et le Hasard*. Ernest Flammarion, Paris.

Baker, K. M. (1975) *Condorcet: From Natural Philosophy to Social Mathematics*. Univ. Chicago Press, Chicago.

Bauer, R. K. (1955) Die Lexissche Dispersionstheorie in ihrer Statistischen Methodenlehre insbesondere zur Streuungsanalyse (analysis of variance). *Mitteilungsblatt für mathematische Statistik und ihre Anwendungsgebiete*, **7**, 25–45. Also, in Russian translation, in: Chetverikov (1968), pp. 225–238.

Bayes, T. (1763) An essay towards solving a problem in the doctrine of chances. *Phil. Trans. Roy. Soc. London*, **53**, 376–398. Also in: (1) *Biometrika*, **45** (1958), 293–315; (2) Pearson and Kendall (1970), pp. 134–153.

Bernoulli, D. (1766) Essai d'une nouvelle analyse de la mortalité causée par la petite vériole et les avantages de l'inoculation pour la prévenir. *Histoire de l'Acad. Roy. Sci.*, 1760; Paris, 1766.

Bernoulli, J. (1713) *Ars Conjectandi.* Basileae impensis Thurnisiorum fratrum, Basle.

Bernoulli, N. (1709) *Specimina Artis conjectandi ad quaestiones Juris applicatae.* Basileae impensis Thurnisiorum fratrum, Basle.

Bernstein, S. N. (1945) On the works of P. L. Chebyshev in the theory of probability. [in Russian.] In: Bernstein (1964), pp. 409–433.

(1946a) *Teoriya Veroiatnostei.* (4th ed.) Gostehizdat, Moscow–Leningrad.

(1946b) A theorem inverse to the theorem of Laplace, and its generalization. [in Russian.], In: (1) Bernstein (1946a), pp. 458–464; (2) Bernstein (1964), pp. 448–454.

(1946c) On the limit theorem of probability theory. [in Russian.] *Izvestiya Nauchno-Issled. Inst. Mat. Meh. Tomsk. Gos. Univ.*, **3**, 174–190. Also in: Bernstein (1964), pp. 434–447.

(1964) *Sobranie Sochineniy: Tom IV, Teoriya Veroiatnostei i Matematicheskaia Statistika* [1911–1946]. Nauka, Moscow.

Berthelot, M. (1888) Notice sur les Origines et sur l'Histoire de la Société Philomathique. pp. i–xvii of: *Société Philomathique de Paris. Mémoires publiés par la Société Philomathique à l'Occasion du Centenaire de sa Fondation 1788–1888.* Gauthier-Villars, Paris.

Bertrand, J. (1875) Note relative au théorème de M. Bienaymé. *C. R. Acad. Sci., Paris,* **81**, 458; 491–492.

(1883) [Book Reviews of] *Théories Mathématiques de la Richesse Sociale,* par Léon Walras; *Recherches sur les Principes Mathématiques de la Théorie de la Richesse,* par Augustin Cournot. *Journal des Savants,* 499–508.

(1887) Solution d'un problème. *C. R. Acad. Sci., Paris,* **105**, 369.

(1888a) Sur l'association des électeurs par le sort. *C. R. Acad. Sci., Paris,* **106**, 17–19.

(1888b) Sur l'erreur à craindre dans l'évaluation des trois angles d'un triangle. *C. R. Acad. Sci., Paris,* **106**, 125–128.

(1889) (1907) *Calcul des Probabilités.* Gauthier-Villars, Paris. (2nd ed., of 1907, is available also in a 1973 printing by Chelsea, N.Y.).

Bessel, F. W. (1812–1813) Untersuchungen über die Bahn des Olberschen Kometen. *Abh. Akad. Berlin* (Math. sect.), 117–160.

Bienaymé, A. F. A. (1887) *Les Machines Marines,* Cours Professé à l'Ecole d'Application du Génie Maritime. E. Bernard, Paris. (Ouvrage couronné par l'Académie des Sciences.)

Bikelis, A. (1969) On an estimate of the remainder term in the central limit theorem for samples from finite sets. [in Russian.] *Studia Scientiarum Mathematicarum Hungarica,* **4**, 345–354.

Bilham, E. G. (1926) Correlation coefficients. *Quart. J. Roy. Met. Soc.,* **52**, 172.

Bortkiewicz (Bortkewitsch), L. von (1894–1896) Kritische Betrachtungen zur theo-
retischen Statistik. *Jahrbücher für Nationalökonomie und Statistik*, (3) **8** (1894),
641–680; **10** (1895), 321–360; **11** (1896), 701–705. Also, in Russian translation, in:
Chetverikov (1968), 55–137.

(1898) *Das Gesetz der kleinen Zahlen*. G. Teubner, Leipzig.

(1909) Statistique. *Encyclopédie des Sciences Mathématiques Pures et Appliquées*. (Edition Française.) Tome I, quatrième volume, troisième fasc., 453–480.
Gauthier-Villars, Paris; Teubner, Leipzig.

(1918) Der mittlere Fehler des zum Quadrat erhobenen Divergenzkoeffizienten.
Jber. Deutsch. Math. Verein., **27**, 71–126.

(1922) Das Helmertsche Verteilungsgesetz für die Quadratsumme zufälliger Beo-
bachtungsfehler. *Z. Angew. Math. Mech.*, **2**, 358–375.

(1931) The relations between stability and homogeneity. *Ann. Math. Statist.*, **2**,
1–22.

Bradley, M. (1976) Scientific education for a new society: The Ecole Polytechnique
1795–1830. *History of Education* 5, 11–24.

Bradman, D. G. (1950) *Farewell to Cricket*. Hodder & Stoughton, London.

Browne, T. (1658) *Hydriotaphia, urne buriall, or, a discourse of the sepulchrall
urnes lately found in Norfolk*. London.

Buniakovsky, V. Ia. (1846) *Osnovania Matematicheskoi Teorii Veroiatnostei*. St.
Petersburg.

Burkhardt, H. (1908) Entwicklungen nach oscillierenden Functionen und Integration der Differentialgleichungen der mathematischen Physik. *Jahresbericht der
Deutschen Mathematiker-Vereinigung*. **10,** Zweites Heft. (See: Erster Halbband,
804–823.)

Callot, J. P. (1958) *Histoire de l'École Polytechnique: ses légendes, ses traditions,
sa gloire*. Les Presses Modernes, Paris.

Campbell, R. (1859) On a test for ascertaining whether an observed degree of
uniformity, or the reverse, in tables of statistics is to be looked upon as remarkable. *Phil Mag.* (4th ser), **18**, 359–368.

Candolle, A. de (1873) *Histoire des Sciences et des Savants Depuis Deux Siècles*.
H. Georg, Geneva. (There is also an edition of 1885; and a German edition of
1911, printed by Leipzig Akad. Verlag, edited by W. Ostwald.)

Cauchy, A. L. (1821) *Cours d'analyse de l'École Royale Polytechnique*. 1re partie.
Analyse Algébrique. Paris. Also in: *Oeuvres Complètes d'Augustin Cauchy, 2^{e}
série, Tome 3*.

(1835a) Mémoire sur l'Interpolation. Lithographed. Published in (1) *Journal de
Math. Pures et Appliquées*, **2** (1837), 193–205; (2) *Phil. Mag.*, **8** (1836), 459–468
(in English tr.); (3) *Oeuvres Complètes d'Augustin Cauchy*, 2e série, Tome 2.
Gauthier-Villars, Paris (1958), pp. 5–17.

(1835b) Nouveaux Exercices de Mathématiques (Exercices de Prague). In
Oeuvres Complètes d'Augustin Cauchy, 2e série, Tome 10. Gauthier-Villars,
Paris (1895), p. 275 ff.

(1847a) Mémoire sur la détermination des orbites des planètes et des comètes. *C. R. Acad. Sci., Paris*, **25**, 401. Also in: *Oeuvres Complètes d'Augustin Cauchy*. 1-ère série, **10**. Gauthier-Villars, Paris (1897), pp. 374–389.

(1847b) Application des formules que fournit la nouvelle méthode d'interpolation à la résolution d'un système d'équations linéaires approximatives, et, en particulier, à la correction des éléments de l'orbite d'un astre. *C. R. Acad. Sci., Paris* **25**, 650. Also in: *Oeuvres Complètes d'Augustin Cauchy*, 1-ère série, **10**. Gauthier-Villars, Paris (1897), pp. 412–420.

(1853a) Mémoire sur l'evaluation d'inconnues déterminées par un grand nombre d'équations approximatives du premier degré. *C. R. Acad. Sci., Paris,* **36,** 1114–1122.

(1853b) Mémoire sur l'interpolation, ou Remarques sur les Remarques de M. Jules Bienaymé. *C. R. Acad. Sci., Paris,* **37,** 64–69.

(1853c) Sur la nouvelle méthode d'interpolation comparée à la méthode des moindres carrés. *C. R. Acad. Sci., Páris,* **37,** 100–109.

(1853d) Mémoire sur les coefficients limitateurs ou restrictateurs. *C. R. Acad. Sci., Paris,* **37,** 150–162.

(1853e) Sur les résultats moyens d'observations de même nature, et sur les résultats les plus probables. *C. R. Acad. Sci., Paris,* **37,** 198–206.

(1853f) Sur la probabilité des erreurs qui affectent des résultats moyens d'observations de même nature. *C. R. Acad. Sci., Paris,* **37,** 264–272.

(1853g) Sur la plus grande erreur à craindre dans un résultat moyen, et sur le système de facteurs qui rend cette plus grande erreur un minimum. *C. R. Acad. Sci., Paris,* **37,** 326–334.

(1853h) Mémoire sur les résultats moyens d'un très-grand nombre des observations. *C. R. Acad. Sci., Paris,* **37,** 381–385.

Chang, W.-C. (1976) Statistical theories and sampling practice. In Owen, D. B. (Ed.) *On the History of Statistics and Probability.* Dekker, New York pp. 299–315.

Chapman, J. M., and Chapman, B. (1957) *The Life and Times of Baron Haussman.* Weidenfeld and Nicolson, London.

Châteauneuf, Benoiston de, L. F. (1824) *Considérations sur les enfants trouvés dans les principaux états de l'Europe.* Chez Auteur et Chez Martinet Libraire, Paris. (Mémoire lu à l'Académie Royale des Sciences, dans la Séance de 11 Août 1823). (Includes a report by Duméril and Coquebert-Monbret, read 29 Dec., 1823).

(1830) De la durée de la vie chez le riche et chez le pauvre. *Annales d'Hygiène Publique,* Paris, **3.**

(1841) De la durée de la vie chez les savants et les gens de lettre. *Annales d'Hygiène Publique,* Paris, **25.**

(1845) Mémoire sur la durée des familles nobles de France. *C. R. Acad. Sci. Morales et Politiques,* **7,** 210–240. Reprinted in *Ann. d'Hygiène Publique et de Médecine Legale,* **35** (1846), 27–58 and in second edition form in *Mémoires Acad. Sci. Morales et Politiques,* **5** (1847), 753–794.

Chebotarev, A. S. (1961) From the history of the development of the method of least squares. [in Russian.] *Voprosy Istorii Estestvoznaniia i Techniki*, **11**, 20–28.

Chebyshev (Tchébichef), P. L. (1846) Démonstration élémentaire d'une proposition générale de la théorie des probabilités. *Crelle's J. Reine Angew. Mathematik*, **33**, 259–267. Also in: (1) his *Oeuvres;* (2) Chebyshev (1955).

(1855) Sur les fractions continues. [in Russian.] *Uchenie Zapiski Imp. Akad. Nauk po Pervomu i Tretemu Otdeleniam.* **3**, 636–664. [Translation into French, and footnote, by Bienaymé appears in *Journal de Mathématiques Pures et Appliquées*, **3** (1858), 289–323.] Also in: (1) his *Oeuvres;* (2) Chebyshev (1955).

(1859) Sur l' interpolation par la méthode des moindres carrés. *Mém. Acad. Sci. St. Pétersburg*, (8) **1**, No 15, 1–24. Also in Chebyshev (1955).

(1867) Des valeurs moyennes. *Liouville's J. Math. Pures Appl.*, (2) **12**, 177–184 [transl. into French by N. Hanikov; published simultaneously in Russian in *Mat. Sbornik*, (2) **2**, 1–9]. Also in: (1) his *Oeuvres;* (2) Chebyshev (1955).

(1874) Sur les valeurs limites des intégrales. *Liouville's J. Math. Pures Appl.*, (2) **19**, 157–160. Also in his *Oeuvres*.

(1887) Sur deux théorèmes relatifs aux probabilités. Supplement to: *Zapiski Imp. Akad. Nauk* (S.P.-B.), **55**, No. 6. Also in: (1) *Acta Math.*, **14** (1890–1891), 305–315; (2) his *Oeuvres;* (3) Chebyshev (1955).

(no date) *Oeuvres*. A. Markov and N. Sonin (Eds.). 2 Vols. Chelsea, New York.

(1951) *Polnoe Sobranie Sochineniy*, Vol. 5: *Prochie Sochinenia, Biograficheskie Materialy*. AN SSSR, Moscow-Leningrad.

(1955) *Izbrannie Trudy*. AN SSSR, Moscow.

Chetverikov, N. S. (Ed.)(1968) *O Teorii Dispersii*. Statistika, Moscow.

Chetverikov, S. S. (1926) On certain aspects of the evolutionary process from the standpoint of modern genetics. [in Russian.] *Zhurnal Eksperimentalnoi Biologii* **A2**, 3–54. [Also, in English translation, in: *Proc. Amer. Phil. Soc.*, **105** (1961), 167–195. N. S. Chetverikov is a brother.]

Chrystal, G. (1952) *Algebra*, Part II (6th ed.). Chelsea, New York.

Chuprov (Tschuprow), A. A. (1905) Die Aufgaben der Theorie der Statistik, *Schmoller's Jahrbuch für Gesetzgebung, Verwaltung und Volkswirtschaft im Deutschen Reich*, **29**, 421–480.

(1910) *Ocherki po Teorii Statistiki*. (1st ed., 1909). St. Petersburg. (Exists in reprinted form, Moscow, 1959).

(1916) On the mathematical expectation of the coefficient of dispersion. [in Russian.] *Izvestiya Akad. Nauk.*, S.P.-B., **10**, 1789–1798.

(1918–1919) Zur Theorie der Stabilität statistischer Reihen. *Skandinavisk Aktuarietidskrift*, **1**, 199–256. Also in: Chetverikov (1968), 138–224.

(1922) Ist die normale Stabilität empirisch nachweisbar? *Nordisk Statisk Tidskrift*, **1**, 369–393.

(1960) *Izbrannie Stati. Voprosy Statistiki*. Moscow.

Condorcet, Le Marquis de (1785) *Essai sur l'application de l'analyse à la probabilité des décisions rendues à la pluralité des voix.* Imprimerie Royale, Paris.

Cournot, A. A. (1838) Mémoire sur les applications du calcul des chances à la statistique judiciaire. *Liouville's J. Math. Pures Appl.*, (1) **3**, 257–334.

(1843) *Exposition de la théorie des chances et des probabilités.* Hachette, Paris.

Cramér, H. (1946) *Mathematical Methods of Statistics.* Princeton University Press, Princeton.

Crosland, M. (1967) *The Society of Arcueil. A View of French Science at the Time of Napoleon I.* Heinemann, London.

Czuber, E. (1899) Die Entwicklung der Wahrscheinlichkeitstheorie und ihrer Anwendungen. *Jahresbericht der Deutschen Mathematiker-Vereinigung,* **7** (2nd part), 1–279.

(1910) *Wahrscheinlichkeitsrechnung.* Vol. 2. Teubner, Leipzig.

Darmois, G. (1928) *Statistique Mathematique.* G. Doin, Paris.

Darroch, J. N. (1964) On the distribution of the number of successes in independent trials. *Ann. Math. Statist.,* **35**, 1317–1321.

David, F. N. (1955) Dicing and gaming (a note on the history of probability). *Biometrika,* **42**, 1–15. Also in: Pearson and Kendall (1970), pp. 1–17.

(1962) *Games, Gods and Gambling: The Origins and History of Probability and Statistical Ideas from the Earliest Times to the Newtonian Era.* Griffin, London.

(1965) Some notes on Laplace. *Bernoulli Bayes Laplace Anniversary Volume* (J. Neyman and L. LeCam, Eds.), pp. 30–44. Springer-Verlag, Berlin.

Davies, M. (1967) Linear approximation using the criterion of least total deviations. *J. Royal Statist. Soc.,* Ser. B, **29**, 101–109.

Demonferrand (Montferrand, de), F. (1838, 1839) Essai sur les lois de la population et de la mortalité en France. *J. de l'école polytechnique,* **16**, 249–323.

(1839) Lois de la population considérées par rapport aux assurances sur la vie. *C. R. Acad. Sci., Paris,* **9**, 211–212.

Deparcieux, A. (1746) *Essai sur les probabilités de la durée de la vie humaine.* Guérin frères, Paris.

Didion, I. (1855) *Société de prévoyance et de secours mutuels de Metz. Calcul du taux des pensions pour la période de 1850 à 1854.* S. Lamort et F. Blonc, Metz. (Didion wrote similar reports for the periods 1855–1859, 1860–1864, etc.).

Dormoy, E. (1878) *Théorie Mathématique des Assurances sur la Vie.* Gauthier-Villars, Paris.

Doubleday, T. (1837) A letter to the Right Honourable Lord Brougham and Vaux, *Blackwood's Edinburgh Magazine,* March 1837, 363–374.

(1842) *The True Law of Population shewn to be Connected with the Food of the People.* Simpkin, Marshall and Co., London.

Dugué, D. (1968) Bienaymé, Jules. *International Encyclopedia of the Social Sciences,* **2**, 73–74. Macmillan and The Free Press, Chicago.

Dupin, C. (1849) Nouvelles recherches sur la population française. *C. R. Acad. Sci., Paris,* **29**, 369–375.

Duvillard (de Durand), E. E. (1806) *Analyse et tableaux de l'influence de la petite vériole sur la mortalité à chaque âge et de cette qu'un préservatif tel que la vaccine peut avoir sur la population et la longévité*. Impr. Impériale, Paris.

Eden, Sir F. M. (1800) *An Estimate of the Number of Inhabitants in Great Britain and Ireland*. J. Wright, London.

Eisenhart, C. (1964) The meaning of "least" in least squares. *J. Washington Acad. Sci.*, **54**, 24–33.

Erlang, A. K. (1929) Opgave Nr. 15. *Mat. Tidsskrift* B, 36.

Fahlbeck, P. E. (1898, 1902) *Sveriges Adel*. (2 volumes) (German translation, Jena 1903).

Faye, H. A. E. A. (1875) *Théorie des Erreurs*. Paris. [Available also in authorized Spanish translation (1888) as *Teoria de los Errores*. México. Imprenta del gobierno. 55 pp.].

Feller, J. (1961) Cournot, (Antoine Augustin) *Dictionnaire de Biographie Française*, **9**, 983–984. Letouzey et Ané, Paris.

Finlaison, J. (1829) *Report of John Finlaison, Actuary of the National Debt, on the Evidence and Elementary Facts on which the Tables of Life Annuities are founded*. Ordered, by the House of Commons, to be printed, 31 March, 1829.

Fisher, R. A. (1920) A mathematical examination of the methods of determining the accuracy of an observation by the mean error, and by the mean square error. *Monthly Notices Roy. Astron. Soc.*, **80**, 758–770. [Reprinted in Fisher's *Contributions to Mathematical Statistics*. Wiley, New York (1950), 2.757a–2.770].

(1922) On the mathematical foundations of theoretical statistics. *Phil. Trans. Roy. Soc.* (Series A), **222**, 309–368. [Reprinted in Fisher's *Contributions to Mathematical Statistics*. Wiley, New York (1950), 10.308a–10.368].

(1925) Theory of statistical estimation. *Proc. Camb. Phil. Soc.*, **22**, 700–725. [Reprinted in Fisher's *Contributions to Mathematical Statistics*. Wiley, New York. (1950), 11.699a–11.725].

(1928) On a distribution yielding the error functions of several well known statistics. *Proc. Int. Congr. Math.*, (Toronto)(1924), pp. 805–813.

(1958) *Statistical Methods for Research Workers*. (13th ed.; 1st ed.: 1925). Oliver and Boyd, Edinburgh.

Fisher, W. D. (1961) A note on curve fitting with minimum deviations by linear programming. *J. Amer. Statist. Assoc.*, **56**, 359–362.

Fisz, M. (1963) *Probability Theory and Mathematical Statistics*. (3rd ed.). J. Wiley, New York.

Florence, D. M. (1739) *Calcul de Jeu appellé par les François le trente-et-quarante, et que l'on nomme à Florence le trente-et-un*. [cited by Todhunter (1865), p. 205].

Fourier, J. B. (1821) Notions générales sur la population. *Recherches statistiques sur la ville de Paris et le département de la Seine*, Vol. I, lx–lxxviii.

(1826) Mémoire sur les résultats moyens déduits d'un grand nombre d'observations. *Recherches statistiques sur la ville de Paris et le département de la Seine*, Vol. III, ix–xxxv.

(1829) Second mémoire sur les résultats moyens et les erreurs des mesures. *Recherches statistiques sur la ville de Paris et le département de la Seine,* Vol. IV, ix–xlviii.

Franceschini, É. (1954) Bienaymé (Irenée-Jules). *Dictionnaire de Biographie Française,* **6,** 415. Letouzey et Ané, Paris.

Freudenthal, H. (1971) Cauchy, Augustin-Louis. *Dictionary of Scientific Biography.* (C. C. Gillispie, Ed.). **3,** 131–148. Scribner's, New York.

Galton, F. (1869) *Hereditary Genius.* (2nd ed.: 1892) (2nd ed. Reprinted in 1962 by Meridian Books, Cleveland).

(1873) Problem 4001. *Educational Times,* 1 April, p. 17.

(1889) *Natural Inheritance.* Macmillan, London and New York.

and Watson, H. W. (1874) On the probability of extinction of families. *J. R. Anthropol. Inst.,* **4,** 138–144. [See also Galton (1889), Appendix F.]

Gass, S. I. (1958) *Linear Programming.* (2nd ed.: 1964) McGraw-Hill, New York.

Gatine, A. (1897) Bienaymé. In *École Polytechnique. Livre du Centenaire 1794–1894.* (Vol. 3, pp. 314–316). Gauthier-Villars, Paris (1894–1897).

Gauss, C. F. (1809) *Theoria Motus Corporum Coelestium.* Hamburg, Perthes and Besser. (1810) Disquisitio de elementis ellipticis Palladis. *Comm. Goett.,* I. (1821–1826) Theoria combinationis observationum erroribus minimis obnoxiae. *Comm. Goett.,* I. Also in: Gauss (1964).

(1964) *Abhandlungen zur Methode der kleinsten Quadrate.* (transl. into German of various writings of Gauss, including the preceding, by A. Börsch and P. Simon. First published 1887 in Berlin.) Würzburg.

Geiringer, H. (1942a) A new explanation of non-normal dispersion in the Lexis theory. *Econometrica,* **10,** 53–60.

(1942b) Observations on analysis of variance theory. *Ann. Math. Statist.,* **13,** 350–369.

Gelfand, A. E., and Solomon, H. (1973) A study of Poisson's models for jury verdicts in criminal and civil trials. *J. Amer. Statist. Assoc.,* **68,** 271–278.

(1974) Modelling jury verdicts in the American legal system. *J. Amer. Statist. Assoc.,* **69,** 32–37.

(1975) Analysing the decision-making process of the American jury. *J. Amer. Statist. Assoc.,* **70,** 305–310.

Gillispie, C. C. (Ed. in chief) (no date) *Dictionary of Scientific Biography.* American Council of Learned Societies. Scribner's, New York.

(1972) Probability and politics: Laplace, Condorcet, and Turgot. *Proc. Amer. Philos. Soc.,* **116,** 1–20.

Gini, C. (1955) Sur Quelques Questions Fondamentales de Statistique. *Annales de l'Institut Henri Poincaré,* **14,** 245–364.

(1956) Géneralisations et applications de la théorie de la dispersion. *Metron,* **18,** 1–75.

Glass, D. V. (1973) *Numbering the People: the Eighteenth-century Population Controversy and the Development of Census and Vital Statistics in Britain.* Saxon House, Farnborough.

Gleissberg, W. (1945) Eine Aufgabe der Kombinatorik und Wahrscheinlichkeitsrechnung. *Univ. Istanbul Rev. Fac. Sci.* (A), **10**, 25–35.

Gnedenko, B. V. (1948) *Trudy Inst. Istor. Estestvoznan.*, **2** (cited by Maistrov, 1967, 1974).

Godeaux, L. (1973) L'oeuvre mathématique de Adolphe Quetelet (1796–1874). *Janus*, **60**, 97–99.

Goedseels, E. (1900) Étude sur la methode de Tobie Mayer. *Ann. Soc. Sci. Bruxelles*, **24**, 37–58. (rapport par P. Mansion, *ibid.*, 85–88).

(1901) Sur la méthode de Cauchy. *Ann. Soc. Sci. Bruxelles*, **25**, 99–102, 146–149.

(1902) Sur l'application de la méthode de Cauchy aux moindres carrés. *Ann. Soc. Sci. Bruxelles*, **26**, 148–156.

(1909) *Théorie des Erreurs d'Observation.* (3rd edn). Peeters, Louvain; Gauthier-Villars, Paris.

Gouraud, C. (1848) *Histoire du Calcul des Probabilités.* A. Durand, Paris.

Gournerie, J. de la (1878) Lecture de la Note suivante, sur les travaux de *M. Bienaymé. C. R. Acad. Sci., Paris*, **87**, 617–619.

Grand Larousse Encyclopédique (see Larousse, P.)

Grande Dictionnaire Universel du XIXe Siècle. Larousse, Paris (no authors, no date).

Grattan-Guiness, I. (1970) *The Development of the Foundation of Mathematical Analysis from Euler to Riemann.* MIT Press, Cambridge, Mass.

(1972) *Joseph Fourier 1768–1830.* MIT Press, Cambridge, Mass.

Graunt, J. (1662) *Natural and Political Observations mentioned in a following Index, and made upon the Bills of Mortality.* London. (Reprinted in *Natural and Political Observations made upon the Bills of Mortality by John Graunt*, Ed. W. F. Willcox, Baltimore, 1939).

Guibert, A. (1838) Solution d'une question relative à la probabilitié des jugements rendus à une majorité quelconque. *Liouville's J. Math. Pures Appl.*, (1) **3**, 25–30.

Guitton, H. (1968) Cournot, Antoine Augustin. *International Encylcopedia of the Social Sciences*, **3**, 427–430. Macmillan and The Free Press, Chicago.

Hacking, I. (1975) *The Emergence of Probability.* Cambridge University Press, London.

Haldane, J. B. S. (1927) A mathematical theory of natural and artificial selection, V. *Proc. Camb. Phil. Soc.*, **23**, 838–844.

Halley, E. (1693) An estimate of the degrees of the mortality of mankind drawn from curious tables of the births and funerals at the City of Breslau; with an attempt to ascertain the price of annuities upon lives. *Phil. Trans. Roy. Soc.*, **17**, 596–610. Some further considerations on the Breslau bills of mortality. *Phil. Trans. Roy. Soc.*, **17**, 654–656.

Handwörterbuch der Sozialwissenschaften (1959) Fischer, Stuttgart.

Hankins, F. H. (1908) *Adolphe Quetelet as Statistician.* Columbia University Press, New York. (Reprinted 1968.)

Hara, K. (1964) Pascal et l'induction mathématique. *L'Oeuvre Scientifique de Pascal,* Chapter 7. Presses Universitaires de France, Paris.

Hardy, G. H., Littlewood, J. E., and Polya, G. (1967) *Inequalities* (2nd ed.) Cambridge University Press, Cambridge.

Harpe, J. de la (1936) *De l'ordre et du hasard. Le réalisme critique d'A. Cournot.*

Harter, H. L. (1974–1975) The method of least squares and some alternatives. *Int. Statist. Rev.,* Part I, **42** (1974a), 147–174; Part II, **42** (1974b), 235–264; Part III, **43** (1975a), 1–44; Part IV, **43** (1975b), 125–190; Addendum to Part IV, **43** (1975c), 273–278; Part V, **43** (1975d), 269–272.

Heiss, K.-P. (1968) Lexis, Wilhelm. *International Encyclopaedia of the Social Sciences,* **9**, 271–276. Macmillan and The Free Press, Chicago.

Hendricks, F. (1852–1853) Contributions to the history of insurance and the theory of life contingencies. *J. Inst. Actuar.,* **2**, 121–150, 222–258; **3**, 93–120.

(1863) Notes on the early history of tontines. *J. Inst. Actuar.,* **10**, 205–219.

Heyde, C. C., and Seneta, E. (1972) The simple branching process, a turning point test and a fundamental inequality: A historical note on I. J. Bienaymé. *Biometrika,* **59**, 680–683. [Reprinted in Kendall and Plackett (1977), pp. 406–409.]

(1975) Bienaymé. *Bull. Internat. Statist. Inst.* (Proc. 40th Session, Warsaw), **46**, Bk. 2, 318–331.

(1976) Bienaymé, Irenée-Jules. *Dictionary of Scientific Biography,* **15**, Scribner's, New York.

Horváth, R. A. (1967) A statisztika fejlódésc Franciaországban és annak magyar tanulsagai. *Acta Universitatis Szegediensis de Attila József Nominatae. Acta Juridica et Politica,* **14** (Fasc. 4, 126 pp.).

(1973) The centenary of Quetelet's death and the development of statistical discipline. *Bull. Internat. Statist. Inst.* (Proc. 39th Session, Vienna), **45**, Bk. 1, 548–554.

(1975) Statistical ideas of Adam Smith with special regard to Quetelet. *Bull. Internat. Statist. Inst.* (Proc. 40th Session, Warsaw), **46**, Bk. 3, 392–400.

Huygens, C. (1657) *Ratiociniis in Ludo Alea.* In *Exercitionum Mathematicorum,* ed. F. van Schooten, Amsterdam. (The Dutch version is printed with facing French translation in Huygens, C. *Oeuvres complètes* Martinus Nijhoff, The Hague. Vol. 14, 1920.)

Iastremsky (Jastremsky), B. S. (1957) The legend of the miraculous role of the law of large numbers. [in Russian.] *Vestnik Statistiki,* No. 2 (1957), 59–63. [Reprinted in Iastremsky (1964), 140–145.]

(1964) *Izbrannie Trudy.* Gosstatizdat, Moscow.

Imshenetsky, V. G. (1888) Elementary derivation of the law of large numbers in the probability calculus. [in Russian.] *Soobschenia Kharkov. Matem. Obschestva,* (2) **1**, 1–6.

Index Biographique des Membres et Correspondants de L'Academie des Sciences du 22 Décembre 1666 au 15 Novembre 1954. (no authors). Gauthier-Villars, Paris, 1954.

Institute of Actuaries (1869) *The Mortality Experiences of Life Insurance Companies Collected by the Institute of Actuaries.* London.

Isnard, A. N. (1781) *Traité des richesses.* F. Grasset, London and Lausanne.

Isserlis, L. (1927) Note on Chebysheff's Interpolation Formula. *Biometrika, 19,* 87–93.

Jagers, P. (1975) *Branching Processes with Biological Applications.* J. Wiley, London.

John, V. von (1884) Geschichte der Statistik, I. Teil: *Von dem Ursprung der Statistik bis auf Quetelet* (1835). F. Enke, Stuttgart.

Jordan, K. (1972) *Chapters on the Classical Calculus of Probabilities.* Akadémiai Kiado, Budapest.

Kalven, H., and Zeisel, H. (1966) *The American Jury.* Little and Brown, Boston.

Kendall, D. G. (1966) Branching processes since 1873. *J. Lond. Math. Soc., 41,* 385–406. [Reprinted in Kendall and Plackett (1977), pp. 383–404.]

(1975) The genealogy of genealogy: Branching processes before (and after) 1873. *Bull. Lond. Math. Soc., 7,* 225–253.

Kendall, M. G. (1945) On the analysis of oscillatory time-series; appendix: The distribution of runs in certain non-random series. *J. Roy. Statist. Soc., 108,* 125–129.

(1955) *The Advanced Theory of Statistics* Vol. 2. Griffin, London.

and Doig, A. (1968) *Bibliography of Statistical Literature: Vol. 3, pre-1940 with Supplements.* Oliver and Boyd, Edinburgh.

and Plackett, R. L. (Eds.) (1977) *Studies in the History of Probability and Statistics.* Volume II. Griffin, London.

and Stuart, A. (1968) *The Advanced Theory of Statistics* (2nd ed.), Vol. 3. Griffin, London.

Kermack, W. O., and McKendrick, A. G. (1936) Tests for randomness in a series of numerical observations. *Proc. Roy. Soc. Edin., 57,* 228–240.

(1937) Some distributions associated with a randomly arranged set of numbers. *Proc. Roy. Soc. Edin., 57,* 332–376.

Keynes, J. M. (1921) *A Treatise on Probability.* Macmillan, London.

King, A. C., and Read, C. R. (1963) *Pathways to Probability.* Holt, Rinehart and Winston, New York.

Kohn, S. (1930) Chuprov, Alexander Alexandrovich. *Encyclopedia of the Social Sciences, 3,* 462–463. Macmillan, New York.

Kolodziejczyk, S. (1930) La vérification de l'hypothèse sur la constance des probabilités. *Ann. Soc. Polon. Math., 9,* 60–71. [Also in: *C. R. Soc. Sci. Varsovie, 24* (1932), 112–115.]

Koren, J. (Ed.) (1918) *The History of Statistics.* B. Franklin, New York. (Collected and edited by J. Koren for the American Statistical Association. Reprinted 1970.

Contains Faure, F. *The Development and Progress of Statistics in France,* pp. 217–329.)

Lagrange, J. L. (1775) Recherches sur les suites récurrentes dont les termes varient de plusieurs manières . . . différences finies et partielles; et sur l'usage de ces équations dans la théorie des hazards. *Nouveaux Mém. Acad. Sci. Berlin,* 183–272. (Also in: *Oeuvres de Lagrange,* **4,** 151–251, Gauthier-Villars, Paris, 1869).

Lancaster, H. O. (1966) Forerunners of the Pearson χ^2. *Austral. J. Statist.,* **8,** 117–126.

(1968) *A Bibliography of Statistical Bibliographies.* Oliver and Boyd, Edinburgh. [Followed by annual supplements in *Rev. Intern. Statist. Inst.,* = *Int. Statist. Rev.:* **37** (1969), 57–67; **38** (1970), 258–267; **39** (1971), 64–73; **40** (1972), 73–81; **41** (1973), 375–379; **42** (1974), 67–70, 307–311.]

(1969) *The Chi-Squared Distribution.* J. Wiley, New York.

(1970) Problems in the bibliography of statistics. *J. Roy. Statist. Soc.,* (A), **133,** 409–441.

(1972) Development of the notion of statistical dependence. *Math. Chronicle,* **2,** 1–16. [Reprinted in Kendall and Plackett (1977), pp. 293–308.]

Láng, L. Baron (1913) *History of Statistics.* [in Hungarian.] Budapest.

Laplace, P. S. (1774) Mémoire sur la probabilité des causes par les évènements. *Mém. math. phys., Acad. Roy. Sci.,* **6,** 621–656.

(1812) *Théorie Analytique des Probabilités.* [Also 1795 (?); 2nd ed., 1814; 3rd ed., 1820] V. Courcier, Paris. (Also, with all 4 Supplements, 4th added in 1825, in Laplace, 1886).

(1814) *Essai Philosophique des Probabilités.* [As Introduction (pp. i–cvi) to his *Théorie Analytique des Probabilités,* 2nd ed.]

(1818) *Deuxième Supplément à la Théorie Analytique des Probabilités.* In: Laplace (1886), **7,** 531–580.

(1886) *Oeuvres Complètes,* 7. Gauthier-Villars, Paris.

Larousse, P. (publisher) (1960) *Grande Larousse Encyclopédique.* Paris.

Laurent, H. (1873) *Traité du Calcul des Probabilités.* Gauthier-Villars, Paris.

Le Cam, L. (1974) *Notes on Asymptotic Methods in Statistical Decision Theory.* University of Montreal Press, Montreal.

Leibniz, G. W. F. von (1665) *De conditionibus.* Revised for inclusion in *Specimina Juris* (1672). In his *Opera Omnia,* J. Dutens, Ed., Geneva, 1768.

Leprieur, P. (1867) *Répertoire de l'École Impériale Polytechnique* [1854–1863]. Gauthier-Villars, Paris. (A sequel to Marielle, 1855.)

Levasseur, E. (1889) *La Population Française. Histoire de la Population avant 1789 et Démographie de la France comparée à celle des autres Nations au XIXᵉ siècle précédée d'une Introduction sur la Statistique.* 3 Vols. A. Rousseau, Paris.

Levene, H. (1952) On the power function of tests of randomness based on runs up and down. *Ann. Math. Statist.,* **23,** 34–56.

and Wolfowitz, J. (1944) The covariance matrix of runs up and down. *Ann. Math. Statist.,* **15,** 58–69.

Lévy, P. M. G. (1975) L'héritage de Quetelet ou l'illusion mathématique. *Bull. Internat. Statist. Inst.* (Proc. 40th Session, Warsaw), **46**, Bk. 4, 96–100.

Lexis, W. (1876) Das Geschlechtsverhältnis der Geborenen und die Wahrschein-lichkeitsrechnung. *Conrad's Jahrb. für Nat.-Ök. u. Statist.*, **27**, 209–245. [Also in: Lexis (1903), 130–169.]

(1877) *Zur Theorie der Massenerscheinungen in der menschlichen Gesellschaft.* F. Wagner, Freiburg im Breisgau.

(1879) Über die Theorie der Stabilität statistischer Reihen. *Conrad's Jahrb. für Nat.-Ök. u Statist.,* **32**, 60–98. Also in: (1) Lexis (1903), pp. 170–212; (2) Chetverikov (1968), pp. 5–38 (in Russian translation).

(1903) *Abhandlungen zur Theorie des Bevölkerungs und Moral Statistik.* G. Fischer, Jena.

Liapounov (Liapounoff), A. M. (1900) Sur une proposition de la théorie des probabilités. *Izvestia Akad. Nauk. S.P.-B. (Bull. Acad. Sci. St. Pétersbourg)* **13**, 359–386. [Also in: Liapounov (1954), pp. 125–151.]

(1901) Nouvelle forme du théorème sur la limite des probabilités. *Zapiski Akad. Nauk S.P.-B. (Mém. Acad. Sci. St. Pétersbourg)* **12**, 1–24. [Also in: Liapounov (1954), pp. 157–176].

(1954) *Sobranie Sochineniy,* Vol. I. AN SSSR, Moscow.

Libri, G. [Libri-Carrucci Dalla Sommaia, G.] (1834) Rapport sur un Mémoire de Bienaymé sur la *Probabilité des résultats moyens des observations.* (Commis-saires, MM Lacroix, Poisson, Libri rapporteur.) *Procès-Verbaux des Séances de l'Acad. Sci. Paris,* **10**, 533–535.

Lindley, D. V. (1965) *Introduction to Probability and Statistics from a Bayesian Viewpoint. Part 2. Inference.* Cambridge University Press, Cambridge.

Linnik, Yu.V. (1958) *Method of Least Squares and Principles of the Theory of Observations.* Transl. from Russian in 1961 and published by Pergamon, Oxford. (Also as: *Methode der kleinsten Quadrate in moderner Darstellung.* Berlin, 1961).

Liouville, J. (1852) Rapport sur un Mémoire de M. Jules Bienaymé, Inspecteur général des finances, concernant la probabilité des erreurs d'après la méthode des moindres carrés. (Commissaires, MM. Lamé, Chasles, Liouville rapporteur). *C. R. Acad. Sci. Paris,* **34**, 90–92. Also in: *Liouville's J. Math. Pures Appl., (1)* **17**, 31–32.

Loève, M. (1963) *Probability Theory* (3rd ed.) Van Nostrand, Princeton; (1977) (4th ed.). Springer-Verlag, New York.

Löffladt, G. (1971) *Augustin Louis Cauchy rex mathematicorum.* G. Löffladt, Nürnberg.

Lukacs, E. (1960) *Characteristic Functions.* Griffin, London.

Lysenko, T. D. (1951) *The Situation in Biological Science.* (Address delivered at the Session of the Lenin Academy of Agricultural Sciences of the U.S.S.R., July 31, 1948.) Foreign Languages Publishing House, Moscow.

Maistrov, L. E. (1967) *Teoriya Veroiatnostei: Istoricheskii Ocherk.* Nauka, Moscow.

(1974) *Probability Theory. A Historical Sketch.* Translated and Edited by S. Kotz, from Maistrov (1967). Academic Press, New York.

Malthus, T. R. (1798) *An Essay on the Principle of Population.* J. Johnson, London. (Further editions followed, the most comprehensive being the 5th ed. published by J. Murray, London, 1817.)

Mansion, P. (1881) Sur un nouveau principe des probabilités; d'après M. Bienaymé. *Mathesis* (Recueil mathématique publié par P. Mansion et J. Neuberg. Ghent) **1**, 10.

Marielle, C. P. (1855) *Répertoire de l'École Impériale Polytechnique* [1794–1853]. Mallet-Bachelier, Paris.

Markov, A. A. (1884) Proofs of some inequalities of P. L. Chebyshev. [in Russian.] *Soobschenia Kharkov. Matem. Obschestva*, **2**, 105–114. (Also as: Démonstration de certaines inégalites de M. Tchébychef. *Mathem. Ann.*, **24**, 172–180.)

(1898) Sur les racines de l'équation $e^{x^2} d^l e^{-x^2}/dx^l = 0$. *Izvestia Akad. Nauk. S.P.-B. (Bull. Acad. Sci. St. Pétersbourg)*, **9**, 435–446. [Also in: Markov (1951), pp. 254–269.]

(1907) Generalization of the law of large numbers to dependent quantities. [in Russian.] *Izvestia Fiz.-Mat. Obschestva Kazan. Univ.*, **15** (1906), 135–156. [Also in: Markov (1951), pp. 339–361.]

(1912) Reply to P. A. Nekrasov. [in Russian.] *Matematicheskii Sbornik* **28**, 215–227.

(1914) Letter to the Editor. [in Russian.] *Strakhovoe Obozrenie (S.P.-B.)* No. 8 (August, 1914), 558.

(1916) On the coefficient of dispersion. [in Russian.] *Izv. Akad. Nauk. S.P.-B.* (6), **10**, No. 9, 709–718. [Also in: Markov (1951), pp. 523–535.]

(1920) On the coefficient of dispersion (2nd note). [in Russian.] *Izv. Rossiyskoy Akad. Nauk.* (6), **14**, 191–198. [Also in: Markov (1951), pp. 537–547.]

(1924) *Ischislenie Veroiatnostei.* (4th posthumous ed.). Gosizdat, Moscow.

(1951) *Izbrannie Trudy.* AN SSSR, Leningrad.

Medvedev, Zh. A. (1969) *The Rise and Fall of T. D. Lysenko.* Columbia University Press, New York.

Meitzen, A. (1891) *History, Theory and Technique of Statistics.* American Academy of Political and Social Science, Philadelphia. (English translation by R. P. Falkner of the 2 vol. German work: *Geschichte, Theorie und Technik der Statistik.)*

Mentré, F. (1908) *Cournot et la Renaissance du Probabilisme au XIXe Siècle.* Rivière, Paris.

(1927) *Pour qu'on lise Cournot.* Beauchesne, Paris.

Merriman, M. (1877) A list of writings relating to the method of least squares, with historical and critical notes. *Trans. Connecticut Acad. Arts and Sciences*, **4** (1877–1882), 151–232.

Messance (de la Michodière) (1766) *Recherches sur la population des Généralités d'Auvergne, de Lyon, de Rouen, et de quelques provinces et villes du royaume avec des réflexions sur la valeur du bled tant en France qu'en Angleterre depuis 1674 jusqu'en 1764.* Durand, Paris.

Meyer, A. (1874) *Calcul des Probabilités,* publié sur les manuscrits de l'auteur par F. Folie (Posthumous). *Société Royale des Sciences de Liège; Mémoires* (2), **4,** 1–458. (German translation by E. Czuber, Leipzig, 1879.)

Michell, J. (1767) An inquiry into the probable parallax and magnitude of the fixed stars, from the quantity of light which they afford us, and the particular circumstances of their situation. *Phil. Trans. Roy. Soc.,* **57,** 234–264.

Milhaud, G. S. (1927) *Études sur Cournot.* Vrin, Paris.

Mill, J. S. (1875) *A System of Logic Ratiocinative and Inductive,* Vol. 2. (9th ed.). Longmans, Green, Reader and Dyer, London.

Milne, J. (1815) *A Treatise on the Valuation of Annuities and Assurances on Lives and Survivorship.* J. Milne, London.

Mises, R. von (1964) *Mathematical Theory of Probability and Statistics.* (Edited and Complemented by Hilda Geiringer.) Academic Press, New York and London.

Mitrinović, D. S., and Vasić, P. M. (1974) History, variations and generalisations of the Čebyšev inequality and the question of some priorities. *Univ. Beograd., Publ. Elektrotehn. Fak.* (Ser. Mat.-Fiz.), No. 461–497, 1–30.

Moheau (J.-B. A. R. Auget, Baron de Montyon) (1778) *Recherches et considérations sur la population de France.* Montard, Paris. (Reprinted 1912, P. Geuthner, Paris.)

Moivre, A. de (1711) De mensura sortis, seu, de probabilitate eventuum in ludis a casu fortinto pendentibus. *Phil. Trans. Roy. Soc.,* **27,** 213–264. [This memoir was afterwards expanded into Moivre (1718).]

(1718) *The Doctrine of Chances: or, a Method of Calculating the Probabilities of Events in Play.* A. Miller, London (2nd ed., 1738; 3rd ed., 1756).

(1730) *Miscellanea Analytica de Seriebus et Quadraturis.* J. Tomson and J. Watts, London.

Mols, R. (1954) *Introduction à la Démographie Historique des Villes d'Europe du XIVᵉ au XVIIIᵉ siecle.* 3 Vols. Louvain.

Moore, G. H., and Wallis, W. A. (1943) Time series significance tests based on signs of differences. *J. Amer. Statist. Assoc.,* **38,** 153–164.

Nagaev, A. V. (1967) On estimating the expected number of direct descendants of a particle in a branching process. *Theor. Probab. Appl.,* **12,** 314–320.

Netto, E. (1901) *Lehrbuch der Combinatorik.* Teubner, Leipzig.

Nogar, R. J. (1967) Cournot, Antoine Augustin. *New Catholic Encyclopedia,* **4,** 391–392. McGraw-Hill, New York.

Ore, O. (1960) Pascal and the invention of probability theory. *Amer. Math. Monthly,* **67,** 409–419.

Ostrogradsky, M. V. (1838) Extract from a memoir on the probability of judicial errors. [In Russian.] In his: *Polnoe Sobranie Trudov,* **3.** AN Ukr. S.S.R., Kiev, 1961.

Owen, D. B. (Ed.) (1976) *On the History of Statistics and Probability.* Dekker, New York.

Pabst, W. R., Jr. (1974) Statistical Studies of the costs of six-man versus twelve-man juries. *Progress in Statistics.* (J. Gani, K. Sarkadi, and I. Vincze, Eds.) **2**, 601–613. North Holland, Amsterdam.

Pascal, B. (1665) *Traité du Triangle Arithmétique avec Quelques Autres Petits Traitez sur la Mesme Matière.* Desprez, Paris.

(1904–1925) *Oeuvres,* 14 volumes. (L. Brunschvicg, P. Boutroux, and F. Gazier, Eds.) "Les Grands Écrivains de France" series. Hachette, Paris.

(1958) *Pensées.* Texte de l'édition Brunschvicg. Édition précédée de la vie de Pascal par Mme Périer, sa soeur. Introduction et notes par Ch. -M. des Granges. Garnier Frères, Paris. (This edition of *Pensées* is based on that in Pascal's *Oeuvres.* The Port-Royal preedition appeared in 1669, posthumously.)

(1966) *Pensées.* Penguin, Baltimore and Harmondsworth. (Translated, with an Introduction, by A. J. Krailsheimer into English; from French editions of L. Lafuma.)

(1967) *The Provincial Letters.* Penguin, Baltimore and Harmondsworth. (Translated, with an Introduction, into English by A. J. Krailsheimer; written anonymously by Pascal, 1656–1658).

Pasquier, L. G. du (1926) *Le Calcul des Probabilités. (Son Évolution Mathématique et Philosophique.) Hermann, Paris.*

Pearl, R. (1925) Vital statistics of the National Academy of Sciences. *Proc. Natl. Acad. Sci. U.S.A.,* **11**, 752–768.

Pearson, K. (1900) On a criterion that a given system of deviations from the probable in the case of a correlated system of variables is such that it can be reasonably supposed to have arisen from random sampling. *Phil. Mag.,* (5) **50**, 157–175.

(1906) "Das Fehlergesetz und seine Verallgemeinerungen durch Fechner und Pearson". A rejoinder. *Biometrika,* **4**, 169–212.

Pearson, E. S., and Kendall, M. G. (Eds.) (1970) *Studies in the History of Statistics and Probability.* Griffin, London and Hafner, Darien.

Plackett, R. L. (1949) A historical note on the method of least squares. *Biometrika,* **36**, 458–460.

(1972) The discovery of the method of least squares. *Biometrika,* **59**, 239–251. [Reprinted in Kendall and Plackett (1977), 279–291.]

Poggendorf, J. C. (1898) Bienaymé, Irenée Jules. *J. C. Poggendorf's Biographisch Literarisches Handwörterbuch zur Geschichte der Exacten Wissenschaften.* 3 Band (1858–1883) hrsg. von B. W. Feddersen & A. J. v. Oettingen. J. A. Barth, Leipzig, p. 128.

Poinsot, L. (1836) Remarques sur la communication de Poisson. *C. R. Acad. Sci., Paris,* **2**, 398–399. (Poisson's reply appeared on pp. 399–400.)

Poisson, S. D. (1820) Sur l'avantage du banquier au jeu de trente-et-quarante. (1) *Soc. Philomat. Paris Bulletin,* 22–25; (2) *Annal. de Chimie,* **13** (1820), 173–183; (3) *Gergonne's Ann. Math.,* **16**, 173–208.

(1824) Sur la probabilité des résultats moyens des observations. *Connaissance des Temps pour l'an 1827,* 273–302.

(1835) Recherches sur la probabilité des jugements principalement en matière criminelle. *C. R. Acad. Sci., Paris,* **1,** 473–494.

(1836) Note sur le calcul des probabilités. *C. R. Acad. Sci., Paris,* **2,** 395–398.

(1837a) Note sur la proportion des condemnations prononcées par les jurys. *C. R. Acad. Sci., Paris,* **5,** 355–357, 459–463.

(1837b) *Recherches sur la probabilité des jugements en matière criminelle et en matière civile, précédées des règles générales du calcul des probabilités.* Bachelier, Paris.

Polya, G. (1918) Über die Verteilungssysteme der Proportionalwahl. *Zeitschrift für Schweizerische Statistik und Volkswirtschaft,* **54,** 363–387.

(1919) Anschauliche und elementare Darstellung der Lexisschen Dispersionstheorie. *Zeitschrift für Schweizerische Statistik und Volkswirtschaft,* **55,** 121–140.

Price, R. (1771) *Observations on Reversionary Payments and Annuities.* T. Cadell, London. (The fourth and best known edition is dated 1783.)

Quetelet, L. A. J. (1835) *Sur l'Homme et le Développement des ses Facultés, ou Essai de Physique Sociale.* Bachelier, Paris. [*Also* published: (1) by L. Hauman, Brussels, 1836; (2) in English transl. by W. & R. Chambers, Edinburgh, 1842; (3) in Russian transl. by O. N. Bakst, St. Petersburg, 1865. *Reprinted* in 1869 by C. Muquardt, Brussels, as *Physique Sociale: ou, Essai sur le Développement des Facultés de l'Homme.*]

(1844) *Recherches Statistiques.* M. Hayez, Brussels.

(1846) *Lettres sur la Théorie des Probabilités.* (Lettres à S.A.R. le duc régnant de Saxe Coburg et Gotha, sur la théorie des probabilités, appliquée aux sciences morales et politiques.) Brussels, M. Hayez. (Transl. into English by O. G. Downs, published by C. & E. Layton, London, 1849.)

(1848) *Le Système Social et les Lois qui le régissent.* Guillaumin, Paris.

(1852) Sur quelques propriétés curieuses que présentent les résultats d'une série d'observations faites dans la vue de déterminer une constant, lorsque les chances de rencontrer des écarts en plus et en moins sout égales et indépendantes les unes des autres. *Bull. Acad. Royale Belgique,* **19,** Partie 2, 303–317.

(1854) Mémoire sur les variations périodiques et non périodiques de la température d'après les observations faites, pendant vingt ans, à l'Observatoire Royal de Bruxelles. In *Mémoires de l'Académie Royale Belgique* **28.**

Radau, R. (1891) Études sur les formules d'interpolation. *Bulletin Astronomique,* **8,** 273–294, 325–351, 376–393, 425–455.

Rao, C. R. (1965) *Linear Statistical Inference and Its Applications.* J. Wiley, New York.

René, R. (1933) Cournot et l'école mathématique. *Econometrica,* **1,** 13–22.

Reynaud, A. A. L., and Duhamel, J. M. C. (1823) *Problèmes et dévelopments sur diverses parties des mathématiques.* Bachelier, Paris.

Rietz, H. L. (1932) On the Lexis theory and the analysis of variance. *Bull. Amer. Math. Soc.,* **38,** 731–735.

Risser, R. (1932) *Applications de la Statistique à la Démographie et à la Biologie.* Gauthier-Villars, Paris.

Romanovsky, V. (1925) Généralisation d'une inégalité de A. Markoff. *C. R. Acad. Sci., Paris,* **180,** 1468–1470.

(1927) Note on orthogonalizing series of functions and Interpolation, *Biometrika,* **19,** 93–99.

Sagnet, L. (no date) Bienaymé (Irénée-Jules) *Grande Encyclopédie Inventaire Raisonné des Sciences, des Lettres et des Arts* par une Société de Savants et de Gens de Lettres (2e ed.), **6,** 752. H. Lamirault, Paris.

Samuels, S. M. (1965) On the number of successes in independent trials. *Ann. Math. Statist.,* **36,** 1272–1278.

Savage, I. R. (1962) *Bibliography of Non-Parametric Statistics.* Harvard University Press, Cambridge.

Schlömilch, O. (1858) Über Mittelgrössen verschiedener Ordnungen. *Zeitschrift für Math. u. Physik,* **3,** 301–308.

Schrutka, L. von (1941) Eine neue Einteilung der Permutationen, *Math. Annalen,* **118,** 246–250.

Seal, H. L. (1949) Mortality data and the binomial probability law. *Skand. Aktuarietidskr.,* **32,** 188–216.

(1967) The historical development of the Gauss linear model. *Biometrika,* **54,** 1–24. [Reprinted in: Pearson and Kendall (1970), pp. 207–230].

Segond, J. (1911) *Cournot et la Psychologie Vitaliste.* Alcan, Paris.

Seneta, E. (1976) On a contribution of Cauchy to linear regression theory. *Ann. Soc. Sci. Bruxelles,* **90,** 229–235.

Sheynin, O. B. (no date) Anderson, Oskar Johann Viktor. *Dictionary of Scientific Biography.* (C. C. Gillispie, Ed.), **1,** 154–155. Scribner's, New York.

(no date) Bortkiewicz (or Bortkewitsch), Ladislaus (or Vladislav). *Dictionary of Scientific Biography.* (C. C. Gillispie, Ed.). **2,** 318–319. Scribner's, New York.

(1966) Origin of the theory of errors. *Nature,* **211,** 1003–1004.

(1968) On the early history of the law of large numbers. *Biometrika,* **55,** 459–467. [Reprinted with slight modification in: Pearson and Kendall (1970), pp. 231–239.]

(1970) Daniel Bernoulli on the normal law. *Biometrika,* **57,** 99–202. [Reprinted in Kendall and Plackett (1977), pp. 101–104.]

(1971) On the history of some statistical laws of distribution. *Biometrika,* **58,** 234–236. [Reprinted in Kendall and Plackett (1977), pp. 328–330.]

(1972) On the mathematical treatment of observations by L. Euler. *Arch. Hist. Exact Sci.,* **9,** 45–56.

(1973) R. J. Boscovich's work on probability. *Arch. Hist. Exact Sci.,* **9,** 306–324.

(1974) *Teoriya Oshibok P. S. Laplace'a.* Viniti, Moscow. (Contracted version in English translation: "Laplace's theory of errors," *Arch. Hist. Exact Sci.,* **17** (1977) 1–61.

Simpson, T. (1740) *The Nature and Laws of Chance.* E. Cave, London.

Sleshinsky (Sleschinsky, Sleszyński), I. V. (1892) On the theory of the method of least squares. [in Russian.] *Zapiski Mat. Otdelenia Novorossiskago Obschestva Estestvoispitatelei (Odessa)*, **14**, 201–264.

Société de Metz (1862) Note sur la véritable situation de la Société de Prévoyance et de Secours mutuels de Metz; addressée par le Président et les Membres de la Commission du Conseil d'Administration. *C. R. Acad. Sci., Paris*, **54**, 793–796.

Société Philomathique de Paris (1878) *Liste des Membres de la Société Philomathique de Paris. Fondée en 1788*. Soc. Philomath., Paris.

Spengler, J. J. (1965) *French Predecessors of Malthus*. Octagon Books, New York.

Steffensen, J. F. (1930) On Sandsynligheden for at Afkommet uddør. *Matematisk Tidsskrift* B1, 19–23.

Steinmann, J. (1965) *Pascal*. Burns and Oates, London. (Transl. and abridged from 2nd French edition of 1962, by M. Turnell. First French edition, 1954.)

Stephan, F. F. (1948) History of the uses of modern sampling procedures. *J. Amer. Statist. Assoc.*, **43**, 12–39.

Stiefel, E. (1960) Note on Jordan elimination, linear programming and Tchebycheff approximation. *Numerische Math.*, **2**, 1–17.

Stigler, S. M. (1973) Laplace, Fisher, and the discovery of the concept of sufficiency. *Biometrika*, **60**, 439–445. [Reprinted in Kendall and Plackett (1977), pp. 271–277.]

(1974) Cauchy and the witch of Agnesi: An historical note on the Cauchy distribution. *Biometrika*, **61**, 375–380.

(1975a) Napoleonic statistics: The work of Laplace. *Biometrika*, **62**, 503–517.

(1975b) The transition from point to distribution estimation. *Bull. Internat. Statist. Inst.* (Proc. 40th Session, Warsaw), **46**, Bk. 2, 332–340.

Strachey, R. (1892) On the probable effect of the limitation of the number of Ordinary Fellows elected into the Royal Society to fifteen in each year on the eventual number of fellows. *Proc. Roy. Soc.*, **51**, 463–470.

Student (W. S. Gosset) (1908) The probable error of a mean. *Biometrika*, **6**, 1–25.

Takács, L. (1967) *Combinatorial Methods in the Theory of Stochastic Processes*. J. Wiley, New York.

Taton, R. (Ed.) (1961) *General History of the Sciences 3: Science in the Nineteenth Century*. Thames and Hudson, London. (Translation by A. J. Pomerans of: *La Science Contemporaine*, Presses Universitaires de France, 1961.)

Thatcher, A. R. (1957) A note on the early solutions of the problem of duration of play. *Biometrika*, **44**, 515–518. [Also in: Pearson and Kendall (1970), pp. 127–130.]

Todhunter, I. (1865) *A History of the Mathematical Theory of Probability from the Time of Pascal to that of Laplace*. Cambridge University Press, London and Cambridge. (Reprinted in 1949 and 1961 by Chelsea, New York.)

(1869) On the method of least squares. *Trans. Camb. Phil. Soc.*, **11**, 219–238.

Uspensky, J. V. (1937) *Introduction to Mathematical Probability*. McGraw-Hill, New York.

Valson, C. A. (1868) *La Vie et les Travaux de Baron Cauchy*. Gauthier-Villars, Paris.

Vapereau, L. G. (1880) *Dictionnaire Universel des Contemporains*. (2 vols.) 5ᵉ éd. L. Hachette, Paris.

Villarceau, Y. (1849a) (Extrait) Méthode pour calculer les éléments des orbites des planètes, ou, plus généralement, des astres dont les orbites sont peu inclinées à l'écliptique, fondée sur l'emploi des dérivées relatives au temps, des trois premiers ordres de la longitude géocentrique et du prémier ordre de la latitude. *C. R. Acad. Sci., Paris*, **29**, 112–115.

(1849b) Premier mémoire sur les étoiles doubles. *Additions à la Connaissance des Temps pour 1852*, pp. 1–197.

Villermé, L. R. (1843) Rapport verbal sur l'ouvrage de M. Th. Doubleday. *C. R. Acad. Sci. Morales et Politiques*, **4**, 223–242.

Wagner, H. M. (1959) Linear programming techniques for regression analysis. *J. Amer. Statist. Assoc.*, **54**, 206–212.

Walbert, D. F. (1971) The effect of jury size in the probability of conviction: an evaluation of Williams vs. Florida. *Case Western Reserve Law Review*, **22**, 529–554.

Walker, H. M. (1931) *Studies in the History of Statistical Method*. Williams and Wilkins, Baltimore.

Wallis, W. A., and Moore, G. H. (1941a) A significance test for time series analysis. *J. Amer. Statist. Assoc.*, **36**, 401–409.

(1941b) A significance test for time series and other ordered observations. *Natl. Bureau Econ. Res.*, New York. Tech. Paper No. 1.

Watson, H. W. (1873) Solution to Problem 4001. *Educational Times*, 1 August, 115–116.

Weatherburn, C. E. (1947) *A First Course in Mathematical Statistics*. Cambridge University Press, Cambridge.

Westergaard, H. (1932) *Contributions to the History of Statistics*. King, London.

Who's Who Inc. (1968) *World Who's Who in Science From Antiquity to the Present*. Marquis, Chicago.

Witt, J. de (1671) *Waerdye van lyf-renten naer proportie van los-renten*, s'Gravenhage. [Much of the material is translated in F. Hendricks (1852–1853).]

Wölffing, E. (1899) Ergänzung des von E. Czuber in seinem Referat über Wahrscheinlichkeitsrechnung gegebenen Literaturverzeichnisses. *Math. Naturwiss. Verein. Württemberg* (Stuttgart), *Mitteilungen*, (2) **1**, 76–84.

(1901) Nachtrag zu dem Ergänzungsverzeichnis zum E. Czuberschen Bericht über Wahrscheinlichkeitsrechnung. *Math. Naturwiss. Verein. Württemberg* (Stuttgart), *Mitteilungen*, (2) **3B**, 93–95.

Wold, H. (1961) Oskar Anderson, 1887–1960. *Ann. Math. Statist.*, **32**, 651–660.

Wolfowitz, J. (1944) Asymptotic distribution of runs up and down. *Ann. Math. Statist.*, **15**, 163–172.

Wrede, F. J. (1873) Några anmarkningar rörande minsta quadratmethoden. *Öfversigt Förhandl. Acad. Stockholm*, **30**, No. 8, 3–34; No. 10, 21–26.

Zeisel, H. (1971) . . . And then there were none: the diminution of the Federal jury. *Univ. Chicago Law Review*, **38**, 710–724.

Name index*

A

Abbé, E. 56
Adrain, R. 64
Aitken, A. C. 41, 42
Albertsen, K. 117
Ampère, A. M. 18, 108
Anderson, O. 12, 49, 57, 58
André, D. 108, 126, 128
Arago, D. F. 17, 18, 31, 32
Arbuthnot, J. 124
Aristotle 9
Arnauld, Angélique (Mother) 113
Arnauld, A. (Dr.) 114, 115
Auget, J. B. A. R. [Moheau] 22, 24
Augustine of Hippo (St.) 113
Auxiron, C. F. J. 120

B

Bachelier, L. 16
Bacon, R. 62
Baker, K. M. 29
Barbier, J. E. 108
Barlow, D. 129
Bartnieva, L. S. 79
Bauer, R. K. 57, 58
Bayes, T. 3, 26, 42, 97, 99–102
Becquerel, A. H. 18
Berger 130
Bernoulli, D. 4, 25
Bernoulli, Jacob 2, 4, 29, 30, 32, 35, 40–43,
    47, 48, 51, 52, 55, 59, 61, 99, 102, 104,
    113, 114, 122, 136
Bernoulli, Johannes 116
Bernoulli, N. 29
Bernstein, S. N. 58, 60, 102, 112, 122
Berthelot, M. 17, 18
Bertillon, L. A. J. 136, 137
Bertrand, J. L. F. 12, 15, 18, 26, 31, 33,
    48, 67, 71, 108, 125, 126
Bessel, F. W. 125
Bienaymé, A. F. A. 9
Bienaymé, C. P. A. 10
Bienaymé de la Motte, Colonel 9

Bienaymé, G. 10
Bienaymé, L. I. A. 9
Bikelis, A. 60
Bilham, E. G. 127
Biot, J. B. 18
Bonaparte, L. 23
Borel, E. 34
Bortkiewicz, L. von 43–46, 49–53, 56–60
Bossut, C. 115, 116
Bradley, Margaret 6
Bradman, D. G. 105
Browne, T. 119
Buffon, G. L. L. 24
Buniakovsky, V. Ia. 33
Burkhardt, H. 79, 82

C

Callot, J. P. 17
Campbell, R. 60, 61
Candolle, A. de 11, 12, 116, 117
Cardano, G. 2
Castan, A. 135
Cauchy, A. L. 4, 5, 8, 13, 14, 18, 36, 37,
    63, 71–96, 111, 112
Chang, W. C. 23
Chapelas-Coulvier-Gravier 125
Chapman, B. 131
Chapman, J. M. 131
Charles X 6
Chasles, M. 7, 70
Châteauneuf, L. F. Benoiston de 8, 11,
    21, 22, 118, 119, 134, 135
Chebotarev, A. S. 90
Chebyshev, P. L. 5, 8, 9, 13–15, 18, 40, 48,
    58, 77, 80, 81, 87, 90, 94, 96, 97,
    121–124
Chekhov, A. 97
Chetverikov, N. S. 49, 58
Chevreul, M. E. 92
Chrystal, G. 112
Chuprov, A. A. 44, 45, 49, 52, 53, 57–59,
    70
Condorcet, J. A. N. de Caritat de 3, 4, 29,
    30, 33, 34, 108, 115, 125

*Refers to Chapters 1 to 6, and excludes I. J. Bienaymé. We have been unable to trace initials
for some names.

Subject index

Two Companion Series

Studies in the History of Mathematics and Physical Sciences
Edited by M.J. Klein *and* G.J. Toomer

Volume 1

A History of Ancient Mathematical Astronomy
By O. Neugebauer

"A totally different appreciation of Ptolemy is afforded by O. Neugebauer's new three-volume work on early astronomy. The inclusion of the word 'mathematical' is deliberate, for Neugebauer eschews the vague, speculative cosmologies of pre-Socratic philosophers.... for Ptolemy, it is the source par excellence.

Divided into six 'books,' this compendium distills much of a lifetime of scientific research into three volumes, and it is surely one of the landmark publications of this century in the history of astronomy."——*Science*

1975. xxxiii, 1456p. 619 illus. 9 plates. 1 foldout. cloth
(Also available in three separate parts)

Volume 2

A History of Numerical Analysis from the 16th through the 19th Century
By H.H. Goldstine

Virtually all the important and basic methods in numerical analysis were developed in the period which is the subject of this book. The author explains all the consequential methods of the period, among them the emergence of logarithms, the approximate solution of polynomials, numerical integration, trigonometric interpolation, and the numerical integration of differential equations.

1977. xiv, 348p. 30 illus. cloth

Sources in the History of Mathematics and Physical Sciences
Edited by M.J. Klein *and* G.J. Toomer

Diocles: On Burning Mirrors
With Text in Arabic and Greek
English translation and commentary by G.J. Toomer

The first edition of an important text from the most productive period of Greek mathematics will significantly alter previously accepted ideas on the early history of conic sections. This presentation, the first major addition to knowledge of mathematics during the Hellenistic period since Heiberg's work on Archimedes' "Method" in 1907, contains the Greek text of the extracts in Eutocius and the complete medieval Arabic text with English translations using modern notation and extensive commentary.

1976. ix, 249p. (64p. in Arabic, 12p. in Greek) 32 illus. 24 plates. cloth

Encounter With Mathematics
By L. Gårding

1977. ix, 270p. 82 illus. cloth

Designed for students with a good background in high school or some college mathematics, this textbook provides a historical, scientific, and cultural framework for the basic parts of mathematics that are encountered during the four years of college. Nine chapters, ranging from number theory to applications, are devoted to this program. Each chapter starts with a historical introduction, continues with a concise but complete account of some basic facts, and proceeds to examine the present state of affairs, including, if possible, some recent piece of research. Most chapters conclude with one or two passages from historical mathematical papers.

Contents

Models and Reality
Number Theory
Algebra
Geometry and Linear Algebra
Limits, Continuity, and Topology
The Heroic Century
Differentiation
Integration
Series
Probability
Applications
The Sociology, Psychology, and Teaching of Mathematics